Algorithms
Graphs
and
Computers

This is Volume 62 in
MATHEMATICS IN SCIENCE AND ENGINEERING
A series of monographs and textbooks
Edited by RICHARD BELLMAN, *University of Southern California*

A complete list of the books in this series appears at the end of this volume.

Algorithms
Graphs
and
Computers

Richard Bellman
University of Southern California
Los Angeles, California

Kenneth L. Cooke
Pomona College
Claremont, California

Jo Ann Lockett
The RAND Corporation
Santa Monica, California

ACADEMIC PRESS *New York and London*

1970

64705

ACADEMIC PRESS, INC.
111 Fifth Avenue, New York, New York 10003

United Kingdom Edition published by
ACADEMIC PRESS, INC. (LONDON) LTD.
Berkeley Square House, London W1X 6BA

LIBRARY OF CONGRESS CATALOG CARD NUMBER: 77-97484

PRINTED IN THE UNITED STATES OF AMERICA

PREFACE

The development of the electronic computer has profoundly and irrevocably changed the scientific world. In so doing it has simultaneously created numerous opportunities for the application of mathematical ideas and methods to the solution of traditional scientific problems and made possible the exploration of research areas in mathematics and the sciences either previously unattainable or undreamt of. We are, in consequence, living in one of the great times of intellectual history.

Among the most exciting of the new areas of mathematical research and applications are those related to the structure, operation, and control of large systems. Our society is composed of a number of partially interlocking systems, economic, industrial, military, political, educational, and medical. Mathematical problems connected with these systems are thus well worth studying, and indeed are becoming of fundamental importance. The objective of this book is to provide an introduction to these new problems, to some of the mathematical theories and methods employed in their analysis, and to the role of the high speed digital computer in their resolution.

The book is thus addressed to readers seeking a first acquaintance with problems of this type, either for a general view of the methodologies of solution or for specific information concerning mathematical and computational solution methods. It is also addressed to readers seeking to obtain some ideas of the different uses of computers in problem solving. We expect that most readers will have a previous or concurrent course in the elements of computer programming. However, many such courses aim at developing facility with certain specific intricacies of computer programming, rather than an appreciation of the overall power of the computer to aid in the treatment of classes of major problems of science and society. What we hope to develop particularly is skill in problem analysis. We proceed from the original verbal problem, imprecisely formulated, and deliberately so, through mathematical equations to flow charts. Our point of view is that the solution of a problem must begin with precise formulation and conclude with a feasible computational scheme. However, since

v

programming languages vary widely and change quickly we have not, with one or two exceptions, supplied programs. A number of exercises call for the writing of programs, and we expect that many readers will wish to test their skill with some of these.

The principal medium we have chosen to achieve our goals is the mathematical puzzle. Whether or not it is true as McLuhan says that the medium is the message, we feel that our chosen medium is appropriate to our ends, since our puzzles are prototypes of a number of major economic and engineering problems. It is appropriate here to quote the more traditional authority of Leibniz;* "So also the games in themselves merit to be studied and if some penetrating mathematician meditated upon them he would find many important results, for man has never shown more ingenuity than in his plays."

We have systematically applied only two basic mathematical theories throughout the book, dynamic programming and graph theory. In a number of cases, as say in the assignment problem, far superior methods exist. There are several reasons for our concentration upon only two methods. In the first place, we want the student to absorb the idea that there exist systematic methods which can be employed to study large classes of routing, scheduling, and combinatorial problems. In the second place, since we wish the book to proceed at a leisurely pace, space limitations forbid any equally detailed account of other powerful and flexible techniques.

For like reasons, we have been forced to eliminate many other areas of computer application. At the ends of appropriate chapters, we present numerous references to the treatment of problems of the foregoing nature by other theories and by means of special-purpose methods. However, we have made no attempt to be encyclopedic.

Let us turn now to a description of the various chapters. In Chapter One, we consider the problem of traversing a network of streets in such a way as to minimize the time to go from one point to another. Our first objective is to intrigue the curiosity of the reader with this new type of word problem, so close to the traditional word problems of elementary algebra, yet different in that routing is an *optimization* problem. We are allowed a finite number of ways of accomplishing an objective, traversing the network, and we are asked to determine the most efficient way. In this case, we wish to minimize the time spent in traveling. We present two methods for attacking this problem. The first is a method of direct enumeration, which we include first because of its conceptual simplicity and secondly because it enables us to introduce basic ideas about graphs, states, and algorithms. The second is based on converting the optimization problem to the problem of solving a system of equations. These

---

*This passage is taken from the interesting paper by O. Ore, Pascal and the Invention of Probability Theory, *Amer. Math. Monthly* **67** (1960), 409–416.

are similar in form to linear algebraic equations but involve a new arithmetic operation, that of determining the minimum of a finite set of quantities. The theory employed to make this conversion is that of dynamic programming. In fact, the discussion here constitutes a self-contained introduction to dynamic programming. References are given to several books where this versatile theory is presented in greater detail and applied to many other types of problems.

In Chapter Two we concentrate on obtaining a numerical solution by hand or machine calculation. Here we introduce the important method of successive approximations. The digital computer is ideally designed for repetitive techniques and thus for the use of successive approximations. The method is used for the solution of ordinary algebraic equations by various types of iteration, including the Newton-Raphson method. To illustrate the fundamental idea that a knowledge of the geometric or physical structure underlying a problem can suggest important and interesting mathematical methods, we present the Steinhaus method for determining the solution of a system of linear algebraic equations. Finally, the routing equations derived in Chapter One are solved by a technique of successive approximations. What is particularly desirable about the method in this case is that there is convergence to the limit in a finite number of steps.

In Chapter Three we turn to a consideration of the difficulties that arise when we attempt to apply the same methods to large maps. This is an important line to pursue since the problems are similar in nature to those encountered in determining the feasible and optimal modes of operation of large systems. A careful discussion of the formidable barriers encountered in the analysis of large systems is designed to motivate the search for new approaches and new techniques. Our constant aim is to suggest the open-endedness of the mathematical domain we are entering. Also in this chapter we emphasize the fact that the routing problem may be viewed abstractly as that of transforming a system from one state to another at minimum cost of resources. A brief discussion of the significance of this problem in the economic and engineering fields is given.

In Chapter Four we investigate with greater care the existence and uniqueness of the solution of the routing equations obtained in Chapter One. We discuss the concept of "approximation in policy space," and indicate why uniqueness must be carefully ascertained if we are really interested in a numerical solution. The concepts of "upper" and "lower" solutions, of such importance in other branches of mathematics, are utilized in a natural fashion. Since this chapter is mathematically more rigorous than any other, some readers may wish to skim through it rapidly on a first reading. For some, on the other hand, this will provide an easy introduction to some important ideas and techniques.

The remainder of the book is devoted to a study of three other famous puzzles, the wine-pouring problem, the problem of the cannibals and missionaries, and the travelling salesman problem, together with some of their ramifications. The fact that these seemingly disparate problems can be viewed within the same abstract framework will, we feel, do much to enhance the student's appreciation of the unifying power of mathematics. Also we feel that emphasis on word problems develops the ability to formulate questions in mathematical terms, which is to say, to construct mathematical models of physical processes. In the exercises at the end of these chapters we reinforce these ideas with discussions of the Hitchcock–Koopmans–Kantorovich transportation problem, the four-color problem, the determination of optimal play in chess and checkers, and a number of others.

Let us now turn to some suggestions for the use of this book. First, the mathematical prerequisite for reading it is a thorough grounding in high school algebra. Although concepts from calculus are rarely needed in the book, we do use a level of mathematical reasoning and notation with which many students become familiar in a first course in calculus. We also assume that the reader is familiar with the rudiments of matrix theory to the extent of being able to multiply two matrices, but no substantial knowledge of linear algebra is assumed.

Chapters One and Two are prerequisite to understanding later chapters in the book. As we have already stated, Chapter Four may be omitted or skimmed on first reading. The later chapters are independent of one another, except that Chapters Five and Six form a unit on the pouring puzzles. The reader may dip into these to suit his own taste.

Our final suggestion is that the reader take the time to carry through by hand or computer the numerical solution of the examples and exercises in the text, wherever feasible. Only in this way will he achieve a thorough appreciation of the merits and demerits of the techniques discussed, and possibly be led to improvements of his own design. The field is a very active one and none of the methods presented is intended to be definitive.

Finally, we hope that we have communicated some of the fun we have had working on these problems and that the reader will share our enjoyment.

Acknowledgments. We wish to express our gratitude to a large number of students and colleagues who have shared in the preparation of this book by offering suggestions, finding and eliminating obscurities in our writing, and so forth. We cannot name them all, but particular thanks go to Dr. Donald L. Bentley for many helpful ideas, to William Thompson for his assistance in writing part of Chapter 6, and to Donna Beck, Richard Butler, David Huemer, and Donald Platt for reading the manuscript in various stages of its preparation, pointing out numerous errors, and testing some of the exercises. Particular thanks go also to Mrs. A. W. Kelly, Rebecca Karush and Jeanette Blood for their careful typing of several versions of the manuscript.

R. BELLMAN
K. L. COOKE
JO ANN LOCKETT

CONTENTS

Chapter Seven
   CANNIBALS AND MISSIONARIES

Chapter Eight
   **THE 'TRAVELLING SALESMAN' AND OTHER
   SCHEDULING PROBLEMS**

Chapter   One

COMMUTING AND COMPUTING

1. Introduction

In elementary courses in algebra, we study how to convert word problems of various types into problems involving the solution of algebraic equations. Particular favorites are "rate" problems. In these classroom conundrums, people are always busily traversing streets, streams, rivers, highways, byways, and airways by means of a bizarre assortment of transportation devices ranging from jet planes, automobiles, and canoes to bicycles, camels, and horses, and even occasionally by foot. It is usually required to determine a time, a speed or a distance. Sometimes, the rate and route are given and it is the time that is desired; in other cases the time elapsed and the route are the initial data and it is the speed that is missing; in still other situations rate and time are the given information and it is the overall distance that is required.

In elementary calculus courses rate problems of a different type are encountered. In the early stages there are direct problems of determining the rate of change of one variable with respect to another. Then, to illustrate the versatility of the methods, there are applications of these results to the solution of certain kinds of maximization and minimization questions. For example, it may be required to ascertain the route from one point to another that consumes the least amount of time, assuming a path of a specified type.

A typical problem of this nature is the following. A traveller possessing a boat wishes to go from one side of the river at point A to the other side at point B. The river is perfectly straight and of constant width four miles, and the point B is seven miles downstream from the point directly across from A.

1

Supposing that the traveller can row steadily at the rate of three miles per hour and run doggedly at the rate of five miles per hour, where should he land the boat on the other side of the river in order to make the trip from A to B in the minimum time? We suppose that he both rows and runs in a straight line, and that there is no current in the river.

Referring to the diagram below, we see that the problem reduces to finding the value of x in the interval $0 \leq x \leq 7$ which minimizes the expression

$$f(x) = \frac{(16 + x^2)^{\frac{1}{2}}}{3} + \frac{7 - x}{5} \qquad (1)$$

Standard techniques in calculus enable us to solve problems of this nature.

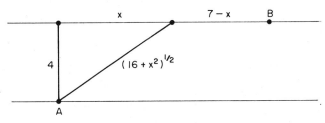

Fig. 1

In this chapter we continue the study of word problems involving travellers and routes, and we are likewise interested in determining paths that require the least time. The difference is, however, that in our case only a *finite* number of possible routes are allowed. This fact, so characteristic of problems that arise in the domains of economics and engineering, makes the use of calculus inappropriate. Entirely new approaches are needed, and as might be suspected new theories have arisen to meet these needs. Our aim here is to present in detail one systematic method for treating questions of this nature, showing how it enables us to transform new types of word problems into new types of equations which are similar in many ways to familiar algebraic equations. In subsequent chapters we discuss how to use a digital computer to solve these equations in numerical terms.

In analyzing these questions in careful detail, our first ambition is to introduce the reader to an important and entertaining class of problems. Our second aim, perhaps of greater long-run import, is to acquaint him in a relatively painless fashion with some of the most powerful and versatile of modern mathematical concepts and methods. In particular, we wish to stress the idea that a "solution" is an algorithm, and to illustrate this basic idea by means of a number of examples.

Specifically, we are interested in algorithms that can be carried out with the aid of a digital computer. This will be discussed at length in Chapters Two and Three.

Our pace throughout will be leisurely. New concepts will always be introduced by means of specific problems and illustrated by means of numerous examples. The authors enjoyed themselves constructing the problems and working out the solutions: we hope that the reader will be as entertained constructing the solutions and formulating new problems. At the end of subsequent chapters appear a number of classical puzzles which can be treated by the methods we are about to describe. References to many other modes of treatment will be found there.

2. A Commuting Problem

We shall begin our discussion with a problem which despite its seeming simplicity will serve to introduce many ideas and methods which have proved to be valuable in a wide range of applications. This is the problem, faced by millions of commuters in the United States, of selecting the best route for driving from home to work and from work to home. How is the choice of a route made?

Some may choose a route because it is shortest in distance; others may prefer the most scenic route. But in our fast-paced present-day civilization, it is likely that most people select the route which is quickest. To do otherwise would be to sacrifice those last few minutes in bed in the morning.

The majority of these wheel-borne commuters experiment with some of the possible routes and then select one which they follow fairly rigidly barring accidents of nature and irresponsibilities of man. It is possible,

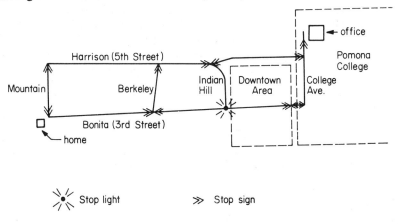

Section of Claremont, California, including campus of Pomona College.

Fig. 2

however, in some cases to determine the quickest route in a quite systematic fashion. Let us see how we might go about this in a particular situation.

In Fig. 2 we see a simplified map of a small section of the city of Claremont, California, including the campus of Pomona College. Let us suppose that a certain systematic professor lives at the corner of Bonita and Mountain Avenues, and that he wishes to find the most rapid way of driving from his residence to his office on the college campus.

To begin with, we observe that the quickest route need not be, and very often isn't, the most direct route: stop signs, traffic lights, congested downtown streets, and school zones must be taken into account.

One way for the methodical professor to begin a solution to his problem is to make a list of *all* possible routes. If he then drives each route in turn and measures the elapsed time for each, he can choose the route of shortest time. This method may be described as a solution by *enumeration of possibilities*. It is simple in concept, and yet fundamental. Indeed, all fundamental ideas are simple in concept. In many cases, as in the map shown in Fig. 2, this method is easily carried out, since there are only a few possible routes. We shall demonstrate this method below.

An additional complication is that the quickest route usually varies with the time of day. This difficulty could be resolved by measuring the time for each route at each time of day at which it is to be driven, and selecting the quickest route for each such time. It will therefore be enough for us to describe how to carry out the solution for any one time of day.*

A more formidable difficulty is that the driving time for a particular route at a particular time may depend on various random, or chance, events. For example, traffic may be heavy one day because of a local store sale or sports event. Since a treatment which would allow for such random events would require the use of the mathematical theory of probability (which we do not assume is familiar to the readers of this book), we shall leave the analytic formulation of problems involving chance events to a more advanced text. References to material of this type will be found at the end of the chapter.

Exercise

1. For use further on, prepare similar maps of routes you frequently or regularly drive: home-to-work, home-to-school, etc.

3. Enumeration of Paths

In order to obtain an idea of how much work is required, let us actually carry through the solution of the quickest route problem for the map

* See the Miscellaneous Exercises in Chapter Three for further discussion.

in Fig. 2, using the straightforward method of enumerating all possibilities mentioned above. Our task is made easier, and the likelihood of overlooking any possibilities is decreased, if the various street intersections are numbered, for example as in Fig. 3. Each possible route can then be described by giving a sequence of digits, beginning with 0 and ending with 8.

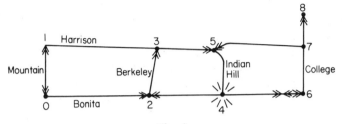

Fig. 3

For example, 0, 2, 4, 5, 7, 8 describes the route which follows Bonita Avenue to Indian Hill Avenue, then follows Indian Hill to Harrison, then Harrison to College Avenue, and finally College Avenue to the office.

With the aid of this shorthand technique, let us now construct a list of all possible routes. We shall, of course, wish to exclude any route which contains the same number twice (a "circuit"). We may return to the same intersection if we take a wrong turn, but in general we do not do it deliberately.*

It is also helpful in achieving a systematic procedure to observe that in any route designation for this particular configuration a given digit can be followed by only a few other digits, not by every other possible digit. For example, referring to Fig. 3, we see that 0 must be followed by 1 or 2; 3 must be followed by 1, 2, or 5; and so on. We thus can use the geometric structure to greatly reduce the number of possible sequences. The possibilities are listed in Table 1.

TABLE 1

0	must be followed by	1 or 2
1	must be followed by	3
2	must be followed by	3 or 4
3	must be followed by	1, 2, or 5
4	must be followed by	2, 5, or 6
5	must be followed by	3, 4, or 7
6	must be followed by	4 or 7
7	must be followed by	5, 6, or 8

* There can be exceptions; see § 18.

In preparing this table, we have observed that neither 1 nor 2 can be followed by 0, since this would create a circuit. In Table 2 we have listed all of the eight possible routes or paths from 0 to 8, after circuits have been removed.

Note also that the intersections labeled 1 and 6 need not have been labeled in Fig. 3, since no branching of paths occurs there. If these were not labeled (but other intersections retained their former labels), the eight routes could be listed as in Table 2.

TABLE 2

0, 2, 4, 6, 7, 8	0, 2, 4, 7, 8
0, 2, 4, 5, 7, 8	0, 2, 4, 5, 7, 8
0, 2, 3, 5, 4, 6, 7, 8	0, 2, 3, 5, 4, 7, 8
0, 2, 3, 5, 7, 8	0, 2, 3, 5, 7, 8
0, 1, 3, 2, 4, 6, 7, 8	0, 3, 2, 4, 7, 8
0, 1, 3, 2, 4, 5, 7, 8	0, 3, 2, 4, 5, 7, 8
0, 1, 3, 5, 4, 6, 7, 8	0, 3, 5, 4, 7, 8
0, 1, 3, 5, 7, 8	0, 3, 5, 7, 8

The reader should convince himself that we could just as well have labelled the intersections A, B, C, D, E, F, G, H, I. We are not using the arithmetic properties of the symbols 0, 1, 2, $\cdots$, 8; merely the fact that they constitute a set of nine distinct symbols. Subsequently, when we consider the use of digital computers, we will see that it becomes important to use numbers rather than other types of symbols for identification.

Exercises

1. Carry out the same procedures for the maps prepared in the previous exercise.

2. What modifications are necessary to take account of the fact that it may be necessary to park several blocks away from the actual destination? What is the situation if there are several possible parking places? Let us agree to avoid the more difficult problems where the chance of finding a parking place depends on where we attempt to park.

4. Tree Diagrams

We have pointed out in the foregoing section that it is not difficult to overlook possible paths. Let us now discuss a simple systematic procedure to help us in this task of enumeration. As we shall see, this procedure occupies a most important role in connection with the use of modern computers.

The so-called *tree diagram* or *logic tree* is an efficient device for enumerating all possibilities of combinations of different events and situations. For example, suppose that we have an urn with three balls in it, numbered 1, 2, and 3, and that we draw them out one at a time until we have drawn all three. How many different orders are there in which we could draw the balls?

To answer this question, let us examine the process. At the first draw we can, of course, draw the ball numbered 1, or the one numbered 2, or the 3. These possibilities can be depicted thus:

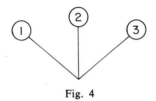

Fig. 4

At the second draw, we can obviously draw either of the two remaining balls. The diagram which follows shows all these possibilities. Finally, we draw the remaining ball.

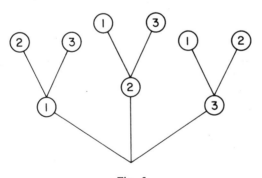

Fig. 5

Figure 6 shows all possible draws. A diagram such as this is often called a *tree* or *logic tree*, because of its resemblance to a real tree, or at least to the branches of a tree. Each path from the start to the end (bottom to top) represents one possible order of events, one logical possibility. In the foregoing example, there are six possible orders in which the balls can be drawn, since the tree has six *branches*. As we see from Fig. 6, these orders can be listed as:

| 1, 2, 3 | 2, 3, 1 | 3, 1, 2 |
| 1, 3, 2 | 2, 1, 3 | 3, 2, 1 |

Observe that here all orders were possible.

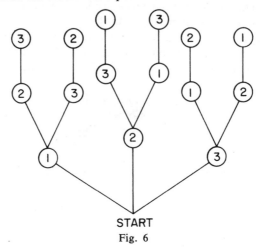

START

Fig. 6

The same technique can be used to assist in the enumeration of routes which resulted in Table 2. Let us attempt to "grow a tree" of possibilities. In Fig. 3, we begin at the intersection labeled 0 and go either to 1 or 2. Therefore our tree begins as follows:

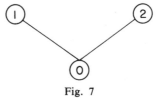

Fig. 7

Using Table 1 or Fig. 3, we see that 1 can be followed only by 3, and 2 by 3 or 4. At this stage, all allowable possibilities are shown on the tree which follows:

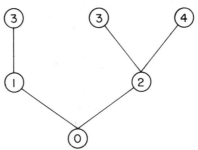

Fig. 8

Now in accordance with Table 1, 3 can be followed by 1, 2, or 5, and 2, 5, or 6. However, as pointed out before, we exclude some of these ssibilities because they correspond to a return to an intersection already visited, that is to say, they lead to a circuit. This is indicated in the following diagram, where dotted circles and lines indicate rejected possibilities.

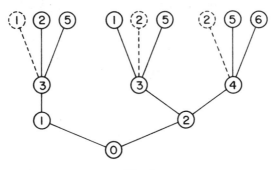

Fig. 9

The complete tree is shown in Fig. 10, which includes all rejected routes. Each path through the tree from 0 to 8 along a solid line represents a possible route. It is seen that there are eight of these, the same eight listed in Table 2. As before, dotted circles and lines represent rejected possibilities. In practice, it is generally not necessary to include these possibilities on the diagram.

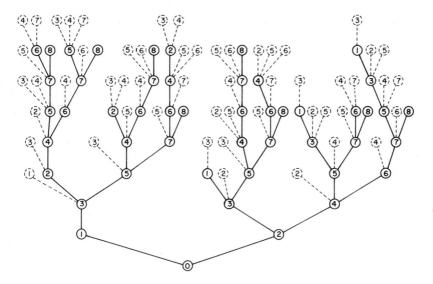

Fig. 10

It should be kept clearly in mind that the logic tree is not a map. It does not show the physical or real location of places or streets. It is a *diagram*, the purpose of which is to help us to keep track of a succession of events, or a set of possibilities. It can be considered to represent a set of possible maps in time, as opposed to the usual map in space.

The reader with some experience in the use of a digital computer, particularly in programming, will recognize an analogy between tree diagrams and *flow charts*. A flow chart is, in fact, a diagram or graph which indicates the flow of events in the computer. All possible sequences of events are indicated on a flow chart, just as in our logic tree above. We shall return to this point below when we draw flow charts for the numerical solution of some of our problems.

Exercises

1. Construct the trees corresponding to the maps drawn in the Exercise of § 2.

2. An urn contains three black balls and two white balls. We draw out three balls, one at a time, leaving two balls in the urn. Draw a tree diagram to depict this experiment.

5.   Timing and the Shortest Route

Once the possible routes have been enumerated, we must determine the length of time required to traverse each one. It is not our intention to enter into the details of how these measurements can be made, but we shall point out some pertinent factors. There are at least two experimental procedures which can be employed. First, our professor can try each of the eight possible routes and record the time for each run, as we have already suggested.* Alternatively, since each route is made up of a number of parts, namely the sections between street intersections (which we shall call *edges* or *links*), he can drive each allowable link and measure the time to traverse it, which we call the *link time*. For example, he might find that the link from 0 to 1 takes 28 seconds, the link from 1 to 3 takes 46 seconds, and so on. The time for each of the eight routes can then be found by adding the times for its separate links.

When we come to larger maps, there may be far more routes than individual links, and an experiment of the first kind may be prohibitively

---

* If account is to be taken of random fluctuations, a number of runs will have to be made over each route. From the times for these runs, the mean time and other statistical data for each route can be estimated. As we have stated, we shall avoid any discussion of these statistical matters.

more expensive than one of the second kind. This is illustrated in the next section.

On the other hand, a disadvantage of the latter procedure is that the time required for a given link may vary somewhat depending on how we approach it—for example, on whether we enter it on a straight path, by a turn, or after a stop sign and turn. In some cases, these variations may be so small that we can afford to neglect them, but in others it may be necessary to take them into account. We shall later describe a method for doing this,* but for the present let us ignore these variations. We want to obtain improved mathematical techniques before we attempt more complicated problems.

We do wish to alert the reader to the fact that the operational solution of actual routing problems may involve a number of factors not even hinted at in an antiseptic mathematical version. The real world is seldom as neat and tidy as textbook examples would give us to believe.

For the purposes of our example we have computed some estimated times by the second method, and have recorded them in Table 3. The times listed are based on assuming an average speed of 30 mph on fast streets, 20 mph on other residential or somewhat congested streets, and

TABLE 3

TABLE OF TIMES FOR FIG. 3

Edge	Distance (feet)	Average speed (feet per sec)	Time (sec)	Delays at stops	Total time (sec)
0,2	1750	44	40	10	50
0,1	800	44	18	10	28
1,3	2000	44	46	0	46
2,3	750	44	17	10	27
2,4	1200	44	27	25	52
3,1	Need not be considered				
3,2	Same as 2,3				
3,5	750	44	17	30	47
4,2	Need not be considered				
4,5	700	44	16	0	16
4,6	1250	15	83	20	103
5,3	Need not be considered				
5,4	700	44	16	10	26
5,7	1500	29	52	10	62
6,7	700	15	47	0	47
7,8	500	15	33	10	43

* See § 18.

TABLE 4

ROUTE TIMES FOR FIG. 3

Route	Total time (sec)
0, 2, 4, 6, 7, 8	295
0, 2, 4, 5, 7, 8	223
0, 2, 3, 5, 4, 6, 7, 8	343
0, 2, 3, 5, 7, 8	229
0, 1, 3, 2, 4, 6, 7, 8	346
0, 1, 3, 2, 4, 5, 7, 8	274
0, 1, 3, 5, 4, 6, 7, 8	340
0, 1, 3, 5, 7, 8	226

10 mph on heavily congested business streets (e.g., Bonita Avenue in down-town area) and in the college area, where there is very heavy pedestrian traffic. In addition, an average delay of 10, 20, or 30 seconds is assumed for each stop sign, depending on how long one must expect to wait for a break in traffic. The stop light is found to cause an average delay of 25 or 10 seconds, depending on whether one is traveling on Bonita or on Indian Hill.

Using the data in Table 3 or the data obtained from a suitable experiment, we can easily find the total time for each of the eight possible routes. The results deduced from Table 3 are given in Table 4. Evidently, 0, 2, 4, 5, 7, 8 is the best route. That is to say, the quickest route is via Bonita to Indian Hill, up to Harrison, over to College, and up. Observe that the route of least distance, 0, 2, 3, 5, 7, 8, is in this case the third fastest route.

Exercise

1. Determine the best routes in the home–school journey, etc.

6. An Improved Map

Although the method of enumeration of all possibilities is satisfactory for the simple map of Fig. 2, it becomes rather unwieldy in more realistic situations. To be convinced of this, let us look at a better map of Clare-mont—but only a slightly better map—as in Fig. 11 below. On this figure we have already numbered the street intersections, and we see that there are not many more than there were before. Proceeding exactly as before, we first compile Table 5, which shows where one can go from each inter-section. Table 5 is scarcely more extensive than Table 1, but when we try to form a list of all possible routes, we find that there are some 38 possi-

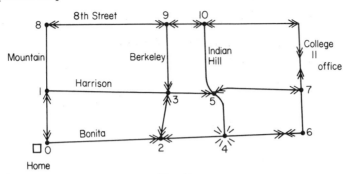

Fig. 11

bilities, instead of the eight in Table 2. Merely to list them all, being sure not to overlook any, has now become somewhat trying, but the reader is urged to complete this task and to compare his results with Table 6. Furthermore, we suggest that he construct an appropriate tree diagram.

TABLE 5

0	must be followed by	1 or 2
1	must be followed by	3 or 8
2	must be followed by	3 or 4
3	must be followed by	1, 2, 5, or 9
4	must be followed by	2, 5, or 6
5	must be followed by	3, 4, 7, or 10
6	must be followed by	4 or 7
7	must be followed by	5, 6, or 11
8	must be followed by	1 or 9
9	must be followed by	3, 8, or 10
10	must be followed by	5, 9, or 11

Although the use of a more complete map has raised the number of possible routes from 8 to 38, the total number of individual sections of street which can be parts of a route has not even been doubled. Consequently, even in this simple case it is far more efficient to determine by experiment the times associated with the links rather than with the set of complete routes. This will be increasingly the case as more streets are added to our map. In Table 7, we list theoretical times for all possible links in a route, computed in the same way as in Table 3. The number of these has increased from 15 to 30, but the rate of increase is not as rapid as the increase in total number of routes.

All that now remains to be done is to find the total time required to traverse each of the routes, and then to pick out the smallest of these. We leave it as an exercise for the reader to complete this task and to compare

TABLE 6

ENUMERATION OF PATHS AND TIMES FOR FIG. 11

	Times
0, 1, 3, 2, 4, 5, 7, 11	274
0, 1, 3, 2, 4, 5, 10, 11	280
0, 1, 3, 2, 4, 6, 7, 5, 10, 11	496
0, 1, 3, 2, 4, 6, 7, 11	346
0, 1, 3, 5, 7, 11	226
0, 1, 3, 5, 4, 6, 7, 11	340
0, 1, 3, 5, 10, 11	232
0, 1, 3, 9, 10, 11	225
0, 1, 3, 9, 10, 5, 7, 11	283
0, 1, 3, 9, 10, 5, 4, 6, 7, 11	397
0, 1, 8, 9, 3, 2, 4, 5, 7, 11	355
0, 1, 8, 9, 3, 2, 4, 5, 10, 11	361
0, 1, 8, 9, 3, 2, 4, 6, 7, 5, 10, 11	577
0, 1, 8, 9, 3, 2, 4, 6, 7, 11	427
0, 1, 8, 9, 3, 5, 4, 6, 7, 11	421
0, 1, 8, 9, 3, 5, 7, 11	307
0, 1, 8, 9, 3, 5, 10, 11	313
0, 1, 8, 9, 10, 5, 3, 2, 4, 6, 7, 11	462
0, 1, 8, 9, 10, 5, 4, 6, 7, 11	392
0, 1, 8, 9, 10, 5, 7, 11	278
0, 1, 8, 9, 10, 11	220
0, 2, 3, 1, 8, 9, 10, 11	315
0, 2, 3, 1, 8, 9, 10, 5, 7, 11	373
0, 2, 3, 1, 8, 9, 10, 5, 4, 6, 7, 11	487
0, 2, 3, 5, 4, 6, 7, 11	343
0, 2, 3, 5, 7, 11	229
0, 2, 3, 5, 10, 11	235
0, 2, 3, 9, 10, 5, 4, 6, 7, 11	400
0, 2, 3, 9, 10, 5, 7, 11	286
0, 2, 3, 9, 10, 11	228
0, 2, 4, 5, 3, 1, 8, 9, 10, 11	373
0, 2, 4, 5, 3, 9, 10, 11	286
0, 2, 4, 5, 7, 11	223
0, 2, 4, 5, 10, 11	229
0, 2, 4, 6, 7, 5, 3, 1, 8, 9, 10, 11	589
0, 2, 4, 6, 7, 5, 3, 9, 10, 11	502
0, 2, 4, 6, 7, 5, 10, 11	445
0, 2, 4, 6, 7, 11	295

TABLE 7

<small>TABLE OF TIMES FOR FIG. 11</small>

Link	Distance (feet)	Average speed (feet per sec)	Time (sec)	Delays at stops	Total time (sec)
0,2	1750	44	40	10	50
0,1	800	44	18	10	28
1,8	1000	44	23	0	23
1,3	2000	44	46	0	46
2,3	750	44	17	10	27
2,4	1200	44	27	25	52
3,1	Same as 1, 3				46
3,2	Same as 2, 3				27
3,5	750	44	17	30	47
3,9	1100	29	38	0	38
4,2	1200	44	27	10	37
4,5	700	44	16	0	16
4,6	1250	15	83	20	103
5,3	750	44	17	0	17
5,4	700	44	16	10	26
5,7	1500	29	52	10	62
5,10	1200	44	27	5	32
6,4	1250	15	83	35	118
6,7	700	15	47	0	47
7,5	1500	29	52	30	82
7,6	Same as 6, 7				47
7,11	500	15	33	10	43
8,1	1000	44	23	10	33
8,9	2000	44	46	10	56
9,3	1100	29	38	10	48
9,8	Same as 8, 9				56
9,10	600	44	14	20	34
10,5	Same as 5, 10				32
10,9	Same as 9, 10				34
10,11	2000	29	69	10	79

his results with those in Table 6. Surprisingly, the quickest route turns out to be 0, 1, 8, 9, 10, 11.

It is interesting to give an estimate of how much arithmetic is required to compute the total time for each path after the table of times is completed. Since there are 38 paths with an average of about eight links each, we see that approximately $38 \times 7 = 266$ additions (of two numbers each) must be performed in order to find the times for all paths. In contrast, for the simpler map in § 3 there are only eight paths, with an average of six links, resulting in about 40 additions. Evidently, the amount of arithmetic has increased even more rapidly than the number of routes, since the routes are now longer. Considerations of this type are vital to the use of digital computers. They are called *feasibility considerations* and are of importance in deciding whether or not we accept a proposed algorithm as an actual solution.

7. The Array of Times

The data on times required to traverse a given link of street, given in Table 3, can be displayed visually in a more helpful pattern. We list the numbers of the intersections 0, 1, 2, 3, 4, 5, 6, 7, 8, along the top and also along the left side of a new table. Then we enter in a given row and column in the table the time required to go *from* the intersection with the number at the left end of the row *to* the intersection with number at the top of the column.* The result is shown in Table 8. For example, the table shows that 16 seconds are needed to drive from intersection 4 to intersection 5, that is, from Bonita to Harrison on Indian Hill.

TABLE 8

MATRIX OF TIMES FOR FIG. 3

To From	0	1	2	3	4	5	6	7	8
0	0	28	50						
1		0		46					
2			0	27	52				
3			27	0		47			
4					0	16	103		
5					26	0		62	
6							0	47	
7								0	43
8									0

The blank spaces in Table 8 indicate that there is no direct connection between the given intersections. For example, the table shows that we can go from 0 to 0 in 0 seconds, from 0 to 1 in 28 seconds, and from 0 to 2

* Of course, we could do it the other way, putting at the top the intersection from which we start, and on the left the intersection to which we move. On all our tables we shall clearly indicate which way we have chosen.

in 50 seconds, but since there are no other entries to the right, no other intersection can be reached directly from intersection 0. Of course, it might be argued that intersection 5, for example, can be reached from intersection 0, by first passing through intermediate intersections, and that a time should therefore be entered in the table opposite the 0 and under the 5. Our convention, however, is to enter times only when there is a direct link between two intersections.

A table of numbers as in Table 8, in which the data are arranged in a square pattern instead of being listed in a single column, is often called an *array* or *matrix* of numbers. The arrangement is a very handy one for quickly reading off the required information. More than that, it has surprising mathematical possibilities as indicated below. Matrices possess their own algebra, and even a calculus.

The individual numbers 28, 50, and so on, appearing in the table will be called the *entries* or *elements* of the array. Since we shall continually be dealing with such arrays or matrices, it is convenient to introduce a special symbol to represent the entries. We shall use the symbol t_{ij} (read this "t sub ij") to stand for the time required to go from intersection i to intersection j along a link. Thus

t_{01} is the time to go from 0 to 1

t_{02} is the time to go from 0 to 2

t_{23} is the time to go from 2 to 3

and so on. From the data in Table 8, applying to Fig. 3, we find that $t_{01} = 28$, $t_{02} = 50$, $t_{23} = 27$; t_{05} however is not defined for this given array. We shall continue to use the t_{ij} symbolism for other arrays. Of course, the values of t_{ij} that appear will depend on the particular array under consideration.

Many readers no doubt know that matrices are of widespread use in mathematics and other subjects. For the benefit of readers not familiar with the basic notions involved, let us digress a bit in order to explain some of them. In mathematical parlance, a matrix is a set of numbers arranged in a square or, more generally, rectangular pattern. It is customary to use a symbol such as t_{ij} (or a_{ij}, etc.) to denote a typical element of a matrix. Thus, the set of quantities

$$t_{11} \qquad t_{12} \qquad t_{13}$$
$$t_{21} \qquad t_{22} \qquad t_{23}$$

is a matrix. The first subscript i of t_{ij} refers to the *row* in which the element t_{ij} appears, while the second subscript j refers to the *column*. If the matrix has m rows and n columns, we say that it is an m by n matrix. If $m = n$, we say that it is a square matrix. For the sake of abbreviation the entire matrix is frequently denoted by a symbol (t_{ij}), in which parentheses enclose the symbol for the ijth element.

One place where the reader may have encountered the concept of a matrix previously is in the study of systems of linear algebraic equations. For example, in connection with equations

$$3x - y + 7z = 0$$
$$2x - y + 4z = \tfrac{1}{2}$$
$$x - y + z = 1$$

the square matrix

$$\begin{array}{ccc} 3 & -1 & 7 \\ 2 & -1 & 4 \\ 1 & -1 & 1 \end{array}$$

is usually called the *matrix of coefficients* and the rectangular matrix

$$\begin{array}{cccc} 3 & -1 & 7 & 0 \\ 2 & -1 & 4 & \tfrac{1}{2} \\ 1 & -1 & 1 & 1 \end{array}$$

is often called the *augmented matrix*. It is clear that these matrices provide all the information needed to obtain the solution of the foregoing linear system. As a matter of fact, efficient methods for solving such systems can be based solely on various techniques for manipulating these matrices. These ideas are elaborated in courses on linear algebra and numerical analysis, and we shall not pursue the matter here. We shall develop some further ideas concerning matrices as we have need of them in later chapters.

The reader will notice that our Table 8 does not quite fit the fore-

TABLE 9

MATRIX OF TIMES FOR FIG. 11

To From	0	1	2	3	4	5	6	7	8	9	10	11
0	0	28	50									
1		0		46					23			
2			0	27	52							
3		46	27	0		47				38		
4			37		0	16	103					
5				17	26	0		62		32		
6					118		0	47				
7						82	47	0				43
8		33							0	56		
9				48					56	0	34	
10						32				34	0	79
11												0

going definition of a matrix, because most of the locations in the 9 by 9 array are blank instead of containing numbers. However, there is no reason why we cannot allow blank entries in matrices if it is convenient for our purposes to do so. Later on we shall show how the use of blank entries can be avoided.

Exercises

1. Construct the array of times for Fig. 11, using the data in Table 7. Check your result with Table 9.

2. The map below shows a part of Moscow, taken from the book, *Elementy dinamicheskogo' programminovania* by E. S. Venttsel, "Navka," Moscow, 1964. Assuming that equal distances require equal times to traverse, construct an array of travel times.

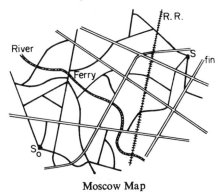

Moscow Map

8. Arbitrary Starting Point

Having carried through the direct enumeration of possible routes for Fig. 3, and again for Fig. 11, we realize that although our method is simple in concept, it can be most tedious in application. Furthermore, we might want more information than we have heretofore asked. For example, we might wish to find the quickest path in Fig. 3 from each possible starting point to the destination 8, or even from each possible starting point to every possible destination. Such a problem might, for example, confront a traffic engineer studying commuter patterns in a whole city, and might be an important preliminary to the development of a master plan for the development of new freeways. To treat this enlarged problem by enumeration, we might go through a procedure identical to the one used previously, for each starting and each stopping point. Since in Fig. 3 any of the nine intersections can be a starting point and any other a stopping point, we find that there are seventy-two problems to be solved, each nearly as hard as the one we have already solved! In Fig. 11, the situation is much worse. Although we can reduce the total amount of work by using parts of one calculation in subsequent calculations, we are nevertheless forced to the conclusion that a better method is desperately needed. In §15, we shall document this statement with some arithmetical calculations.

Enlarging the original problem to allow arbitrary starting and stopping points appears to make the task of solution much longer. However, it actually provides the clue to a new method which will prove to be vastly more efficient than the method of enumeration. We shall elaborate on this observation in a moment, but first let us point out that we shall need an extension of Table 8 or Table 9 before we can attempt the extended problem. Looking at Table 8, we find an entry for "from 0 to 1," but none for "from 1 to 0." For the original problem we would never use this section of street since it would automatically constitute part of a circuit. In our extended problem, however, the starting point may be 1, for example, and it is therefore possible that the quickest path contains the link from 1 to 0. Consequently, we shall have to replace Table 8 by a new table which includes a time for *every* link. The enlarged array is given in Table 10. Notice that the time required to go from i to j is not in every case the same as the time to go from j to i, i. e. that $t_{ij} \neq t_{ji}$ for some i and j. An extreme example of this lack of symmetry occurs in routing problems where there are some one-way streets.

For future reference we give in Table 11 an enlarged array of times for the map in Fig. 11. Times of 33 and 89 seconds have been assigned to the edges from 11 to 7 and 11 to 10, respectively.

TABLE 10

ENLARGED MATRIX OF TIMES FOR FIG. 3

To From	0	1	2	3	4	5	6	7	8
0	0	28	50						
1	28	0		46					
2	40		0	27	52				
3		46	27	0		47			
4			37		0	16	103		
5				17	26	0		62	
6					118		0	47	
7						82	47	0	43
8								33	0

TABLE 11

Enlarged Array of Times for Fig. 11

To From	0	1	2	3	4	5	6	7	8	9	10	11
0	0	28	50									
1	28	0		46					23			
2	40		0	27	52							
3		46	27	0		47				38		
4		37		0	16	103						
5			17	26	0		62				32	
6				118		0	47					
7					82	47	0					43
8		33							0	56		
9				48					56	0	34	
10					32					34	0	79
11							33				89	0

Exercises

1. Enumerate all allowable routes from 1 to 8 in Fig. 3, and find the quickest route.

2. Construct the complete array of times for the maps drawn in the exercise of § 2.

9. Minimum Time Functions

In the previous section, we indicated that it might be expedient to enlarge the original problem to the following more general problem: "Find the least time from any intersection point on the map to the fixed destination."

This idea of imbedding a particular problem within a family of similar problems is one of the most important and powerful in mathematics. It is often the case that the family of problems can be solved simultaneously and elegantly despite the fact that the individual problem remains obdurate. This is a particular illustration of the comparative method that is common to all intellectual disciplines. There are usually many ways of accomplishing this feat of imbedding. The particular method used depends both upon the answers desired and the analytic and computational facilities available.

By enlarging the problem, we have introduced many unknown quantities. Taking our cue from algebra, where one of the devices we learned was that of replacing each unknown quantity by a symbol, we introduce the following quantities,

$$f_0 = \text{the least time in which one can go from}$$
$$\text{intersection 0 to the destination}$$
$$(1)$$
$$f_1 = \text{the least time in which one can go from}$$
$$\text{intersection 1 to the destination}$$

and so on, up to f_7 for Fig. 3 and up to f_{10} for Fig. 11. (We could include f_8 for Fig. 3 and f_{11} for Fig. 11, for they are both equal to zero.) It is convenient to introduce a subscript notation and write

$$f_i = \text{the least time in which one can go from the}$$
$$i\text{th intersection to the destination}$$
$$(2)$$

where i can assume any of the values $0, 1, 2, \cdots, N-1$, and N is the destination ($N = 8$ for Fig. 3 and $N = 11$ for Fig. 11). Here we are using the sophistication of calculus* and introducing a function, f_i. In our case, it is a rather simple function since the argument, i, assumes only the foregoing finite set of values. This introduction of a function is a standard way of generating a family of problems.

In accordance with the precepts of algebra, the next step is to write down a set of equations connecting these unknown quantities, $\{f_i\}$.[†] The final step consists of obtaining the solution of these equations by means of a prescribed rule, or as we shall subsequently say, by means of an *algorithm*. What we mean by "algorithm" and particularly by "feasible algorithm" will be discussed in some detail in what follows. Let us emphasize here that what we accept as an algorithm depends critically upon the available computational facilities. Hence, what we call a "solution" is strongly time dependent. As computers become more powerful, and mathematicians more adroit, some algorithms become candidates for the crown of solution and some get relegated to the museum shelf.

10. The Minimum Function

In order to write these equations for the desired quantities f_i conveniently, it is advisable to introduce a new arithmetic operation on two quantities, the minimum function, written $\min(a, b)$ and defined as follows:

$$\min(a, b) = \text{the smaller of the two real numbers } a \text{ and } b \quad (1)$$

Thus, $\min(5, 3) = 3$, $\min(0, -2) = -2$.

---

* But we make no use of any of the methods or results of calculus.
[†] The symbol $\{f_i\}$ is conventional notation for the sequence of values $f_0, f_1, \cdots, f_N$.

A function, after all, is a set of instructions for obtaining new numbers from old. The function of a and b, min (a, b), may look a bit peculiar as compared to familiar instructions such as $a^2 + b^2$ or $\log a$, but it is nonetheless a perfectly definite and respectable instruction. Furthermore, it is one which can readily be carried out by a computer capable of performing arithmetic. For our purposes, this is an essential characteristic. This function plays an important role in discrete minimization problems, and more generally, in studies of the determination of the efficient operations of systems.

It is easy to see that this function is symmetric, that is

$$\min (a, b) = \min (b, a) \tag{2}$$

In terms of more familiar notation, we note that

$$\min (a, b) = \frac{a + b - |a - b|}{2} \tag{3}$$

where $|a - b|$ denotes, as usual, the absolute value of $a - b$. All of the properties of min (a, b) may thus be deduced from the right-hand side of (3), if we wish.

In similar fashion, let us introduce

$$\min (a, b, c) = \text{the smallest of the three real} \\ \text{numbers, } a, b, \text{ and } c \tag{4}$$

Thus $\min (2, 3, 7) = 2$, $\min (5, 0, -1) = -1$. Once again, it is easy to see that min (a, b, c) is symmetric in its arguments. In other words, the smallest of a set of numbers is independent of the order in which the numbers are given;

$$\min (a, b, c) = \min (a, \min (b, c)) \\ = \min (b, \min (a, c)) \\ = \min (c, \min (a, b)) \tag{5}$$

Generally, if we define min $(a, b, c, \cdots, n)$ in the obvious fashion, we see that

$$\min (a, b, c, \cdots, n) = \min (a, \min (b, c, \cdots, n)) \tag{6}$$

This is an essential structural property.

It is important for both analytic and computational purposes to appreciate what (6) implies as far as a feasible algorithm for computing min $(a, b, c, \cdots, n)$ is concerned. It means that we can use a simple sequential procedure of the following type: Compare a and b, choose the minimum of the two; compare c with this minimum value and choose the minimum of these; and so on.

The introduction of all of these different letters is rather cumbersome.

In more elegant mathematics (which is always simpler mathematics once the language is understood), we write, in place of $a, b, \cdots, n$, the symbols $a_1, a_2, \cdots, a_n$. Then, in place of the expression

$$\min(a_1, a_2, \cdots, a_n) \tag{7}$$

we need merely write

$$\min_{1\le i\le n} a_i \tag{8}$$

Thus, (6) becomes

$$\min_{1\le i\le n} a_i = \min(a_1, \min_{2\le i\le n} a_i) \tag{9}$$

In some cases the symbol in (8) will be replaced by $\min_i a_i$ or merely $\min a_i$ if the range of values of i is clear from the context.

Exercises

1. Let $\max(a, b)$ denote the larger of the two real numbers a and b. Show that $\max(a, b) = (a + b + |a - b|)/2$. Hence, show that

   $$\min(a, b) + \max(a, b) = a + b$$

   Deduce the result from logical considerations alone.

2. Show that $\max_{1\le i\le n} a_i = \max(a_1, \max_{2\le i\le n} a_i)$.

3. Suppose that the quantities a_i first strictly decrease, then strictly increase as $i = 1, 2, \cdots, n$. Is there a more efficient way of finding $\min_i a_i$ than by use of the step-by-step sequential procedure discussed above? As n becomes very large, approximately what percentage of effort can one save? (Hint: If a_i is larger than a_{i+1}, on which side of a_i should we look next for the minimum value?) Problems of this type are part of a modern mathematical theory called "the theory of search." See Chapter 4 of R. Bellman and S. Dreyfus, *Applied Dynamic Programming*, Princeton University Press, 1963.

4. Let $M_2(a_i) =$ the second smallest of the n real numbers $a_1, a_2, \cdots, a_n$. What is the analogue of the relation in (9)?

5. Show that

   $$c + \min(a, b) = \min(c + a, c + b)$$

   and, generally,

   $$c + \min_i a_i = \min_i(a_i + c)$$

6. Under what conditions does

   $$\min_i(ka_i) = k \min_i a_i?$$

7. Show that $|a| = \max(a, -a)$.

8. Show that $\min\left[\min_{1\le i\le k} a_i, \min_{k+1\le i\le n} a_i\right] = \min_{1\le i\le n} a_i$.

9. Show that $\min(a, b, a+c) = \min(a, b)$, if $c \ge 0$.

10. (a) Write a computer program to find and print out $\min_{1\le i\le n} a_i$ if the numbers a_i are stored in the computer memory.

    (b) Given the numbers a_i, $(1 \le i \le n)$ in computer storage, write a program to find $\min a_i$, $\max a_i$ and also the median a_i.

11. Show that $\min(a, \max(b, c)) = \max(\min(a, b), \min(a, c))$.

11. Equations for the f_i

In order to make the new procedure that we shall follow in the general case as clear and plausible as possible, let us consider a very simple case first. Consider the map below.

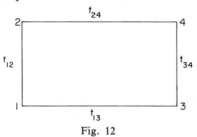

Fig. 12

Let the final destination be 4, and let us seek to determine f_1, f_2, f_3, the quantities representing respectively the quickest times to 4 from intersections 1, 2, and 3. It is clear, upon enumerating possible paths, that

$$f_1 = \min[t_{12} + t_{24}, \ t_{13} + t_{34}]$$
$$f_2 = \min[t_{24}, \ t_{21} + t_{13} + t_{34}] \qquad (1)$$
$$f_3 = \min[t_{34}, \ t_{31} + t_{12} + t_{24}]$$

These are merely the analytic versions of the enumerations we have already considered.

There are, however, less obvious relations which we now wish to exhibit. Let us examine the route of least time from 1 to 4, which we shall call R. Evidently, it begins by going to one of the available next points, either 2 or 3. If the next point on R is 2, the route R thereafter must follow the quickest route from 2 to 4. The proof of this "commonsense" statement is by contradiction: otherwise, there would be a quicker route from 1 to 4 passing through 2.

If, on the other hand, the next point in the path is 3, the route R thereafter must follow the quickest route from 3 to 4.

Thus, we see that f_1 is either $t_{12} + f_2$ or $t_{13} + f_3$. Since f_1 is the *least* time, we have the equation

$$f_1 = \min (t_{12} + f_2, \; t_{13} + f_3) \tag{2}$$

We see now the convenience of the minimum function in writing our equations.

In the same way, we can deduce that

$$f_2 = \min (t_{21} + f_1, \; t_{24})$$
$$f_3 = \min (t_{31} + f_1, \; t_{34}) \tag{3}$$

This reasoning readily applies to any map, not just to the one in Fig. 12. For example, consider Fig. 3 again. Let us define the quantities f_0, $f_1, f_2, f_3, f_4, f_5, f_6, f_7, f_8$, as we already have, that is, for each choice of i, $i = 0, 1, 2, \cdots, 7, 8$,

$$f_i = \text{least time in which one can go from the intersection}$$
$$\text{labelled } i \text{ to the destination, 8} \tag{4}$$

It is obvious that $f_8 = 0$. Also as before, we let t_{ij} denote the time required to traverse the link from i to j. Here i and j are integers from 0 to 8, and t_{ij} is defined when there is a street (link) connecting intersection i to intersection j. To obtain a set of simultaneous equations for the f_i, we proceed as above. Referring to Fig. 3, we see that starting from 0, the quickest route goes either first to 1 or to 2, and thereafter follows the shortest route to 8. The justification for this statement is as before. Therefore,

$$f_0 = \min (t_{01} + f_1, \; t_{02} + f_2) \tag{5}$$

This can be abbreviated to the form

$$f_0 = \min_{j=1,2} (t_{0j} + f_j) \tag{6}$$

where this notation, as before, means that we are to set $j = 1$ and $j = 2$ in succession in the expression in the parentheses, and to choose the smaller quantity.

In the same way, the optimal, or best, route to 8 starting from 2 must go first to 0, 3, or 4 (these being the only points directly accessible from 2), and must then follow the optimal route to 8. It follows that

$$f_2 = \min (t_{20} + f_0, \; t_{23} + f_3, \; t_{24} + f_4) \tag{7}$$

or, in the same abbreviated form as above,

$$f_2 = \min_{j=0,3,4} (t_{2j} + f_j) \tag{8}$$

In a similar fashion, we obtain equations of this nature for $f_3, f_4, \cdots, f_7$. We can condense all of the foregoing equations into one equation,

$$f_i = \min_j (t_{ij} + f_j), \qquad i = 0, 1, \cdots, 7, \qquad (9)$$

where for each i we allow only values of j for which a direct link connecting i and j exists.

We shall discuss a way of avoiding this restriction subsequently in Chapter Two.

12. Reduced Tree

Use of the foregoing equations enables us considerably to simplify the trees of paths given in the previous pages; see Figs. 8, 9, and 10. Thus, Eq. (11.2) is equivalent to:

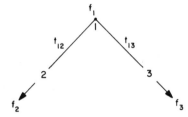

Fig. 13

Equation (11.7) is equivalent to:

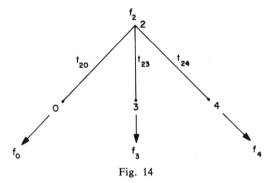

Fig. 14

These reduced trees have been obtained by a systematic application of the following "commonsense" principle: "No matter what point we go to next, the continuation must be a path of quickest time from this new starting point to the fixed destination." As noted above, the proof is by contradiction.

In conceiving of the problem of tracing a shortest route we must make a choice of which point to go to next. The rule that guides our choice, or decision, is the principle stated above.

The fact that we employ the same general rule, or *policy*, at each step has important advantages both with respect to the use of intuition and with respect to analytic and computational efforts. It is of particular significance as far as instructions to a digital computer are concerned.

A further discussion and utilization of the fundamental concept of policy will be found in Chapter Two, particularly § 16 and § 17.

13. General Map

The relations contained in Eq. (11.9) are valid for maps of any type as we have mentioned above. Let us then turn to the general case. Suppose that the map contains N intersection points, labelled $1, 2, \cdots, N$, where N is the destination. As before, t_{ij} denotes the time required to

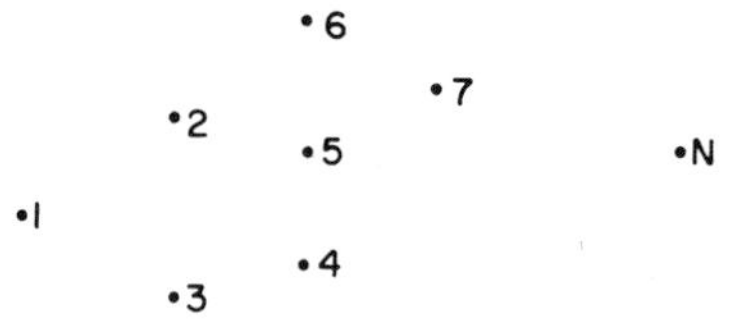

Fig. 15

traverse a direct link between i and j, when it exists, and f_i is the quickest time to get from i to N along some path composed of links. Then the same arguments as before yield the compact equation

$$f_i = \min_j [t_{ij} + f_j], \quad i = 1, 2, \cdots, N - 1, \tag{1}$$

with $f_N = 0$.

In (1), the allowable values of j correspond to intersections which are directly accessible from the intersection i. Since these values of j depend on i in a way which is determined by the particular map under consideration, we cannot indicate the possible values of j explicitly in (1). We shall simply write the equation in (1), or since j must be different from i, the equation

$$f_i = \min_{j \neq i} (t_{ij} + f_j), \quad i = 1, 2, \cdots, N - 1, \tag{2}$$

and shall understand by this convention that j is to range over intersections joined to i by a link.

Let us assume for the moment for the sake of simplicity that every two points are linked and consider once again the justification for (2). Since verbal arguments should always be distrusted, even when persuasively plausible, let us give a simple analytic argument for the validity of (2). The two arguments, the verbal and analytic, are easily seen to be equivalent. However, it is important for the student to learn how to buttress his intuition by means of a formal proof. Occasionally, these procedures disclose, to our consternation, a flaw in an "obvious" statement. It is also essential to learn from experience when intuition can be trusted.

Consider any particular vertex i $(1 \leq i \leq N - 1)$. For any vertex j accessible from i, the route from i to j followed by the quickest route from j to N takes a time $t_{ij} + f_j$. Therefore, the least possible time, f_i, from i to N is certainly no larger than any of the quantities $t_{ij} + f_j$ where j ranges over the values distinct from i. That is,

$$f_i \leq t_{ij} + f_j \quad \text{(all } j \neq i) \tag{3}$$

On the other hand, the quickest route from i to N must start by going to *some* vertex, call it k, accessible from i, and then following *some* route, call it R_{kN}, from k to N. Therefore,

$$f_i = t_{ik} + (\text{time for } R_{kN}) \tag{4}$$

Since, by definition, the time for R_{kN} is no less than f_k, we get

$$f_i \geq t_{ik} + f_k \tag{5}$$

for some k value. Since both (5) and (3) hold for this particular k, it follows that

$$f_i = t_{ik} + f_k \tag{6}$$

Finally, this sum $t_{ik} + f_k$ must be the minimum of all of the quantities $t_{ij} + f_j$, $j \neq i$, since otherwise we should have

$$f_i = t_{ik} + f_k > t_{ij} + f_j \tag{7}$$

for some j, contrary to (3). Thus,

$$f_i = t_{ik} + f_k = \min_{j \neq i} (t_{ij} + f_j) \tag{8}$$

which is precisely (1).

Exercise

1. Suppose that we wish to find the quickest routes from one fixed vertex to every other vertex, instead of the quickest route from each vertex to one fixed terminal vertex. (If $t_{ij} = t_{ji}$ for all i, j, these questions are equivalent.) For a map with N intersection points, let $f_1 = 0$ and for $2 \leq i \leq N$ let f_i denote the minimal time for a route from vertex 1 to vertex i. Show that the numbers f_i satisfy the equations

$$f_i = \min_{j \neq i} (t_{ji} + f_j), \quad 2 \leq i \leq N$$

14. Solving the Equations

For a map with N intersection points, Eq. (13.2) provides N equations in the N unknowns $f_1, f_2 \cdots, f_N$ (assuming, as we do, that the t_{ij} are known quantities). In elementary algebra courses, students usually study sets of simultaneous equations, and are led to believe that they can be solved if the number of unknowns equals the number of equations. This is certainly a useful rule of thumb, although not, as simple examples show, universally true. There are, however, systematic ways of attacking the problem for linear algebraic equations.

It is therefore natural to ask whether there is some equally systematic method for solving the set of Eqs. (13.2), so close in form to traditional algebraic equations. In order to see how this might be done, let us return to the simple map in Fig. 12. There are then only 4 unknowns, f_1, f_2, f_3, f_4, and the four equations are

$$f_1 = \min\left(t_{12} + f_2,\ t_{13} + f_3\right)$$
$$f_2 = \min\left(t_{21} + f_1,\ t_{24} + f_4\right)$$
$$f_3 = \min\left(t_{31} + f_1,\ t_{34} + f_4\right) \tag{1}$$
$$f_4 = 0$$

A well-known general approach to the solution of sets of simultaneous equations is the method of successive elimination of variables. This approach applies here as well. In the first place, the quantity f_4 can obviously be eliminated, in (1) above, leaving the set of three equations in three unknowns

$$f_1 = \min\left(t_{12} + f_2,\ t_{13} + f_3\right)$$
$$f_2 = \min\left(t_{21} + f_1,\ t_{24}\right) \tag{2}$$
$$f_3 = \min\left(t_{31} + f_1,\ t_{34}\right)$$

Next, we see that the second and third equations yield f_2 and f_3 in terms of f_1. Substitution of these relations in the first equation results in the relation

$$f_1 = \min\left[t_{12} + \min\left(t_{21} + f_1,\ t_{24}\right),\ t_{13} + \min\left(t_{31} + f_1,\ t_{34}\right)\right] \tag{3}$$

involving only one unknown, f_1.

This relation looks rather complicated, but it can be readily simplified using the relations

$$\min\left[\min\left(a, b\right),\ \min\left(c, d\right)\right] = \min\left(a, b, c, d\right) \tag{4}$$
$$a + \min\left(b, c\right) = \min\left(a + b,\ a + c\right) \tag{5}$$

(cf., Exercises 5 and 8 at the end of § 10).

Applying (5) to (3), we obtain

$$f_1 = \min\,[\min\,(t_{12} + t_{21} + f_1,\ t_{12} + t_{24}),$$
$$\min\,(t_{13} + t_{31} + f_1,\ t_{13} + t_{34})] \tag{6}$$

Now applying (4), we get

$$f_1 = \min\,(t_{12} + t_{21} + f_1,\ t_{12} + t_{24},\ t_{13} + t_{31} + f_1,\ t_{13} + t_{34}) \tag{7}$$

This is simpler than (3) in that there is only one min symbol, and we have only one equation in one unknown, but unhappily the unknown f_1 still appears on both sides of the equation. Our algebraic training does not at first glance seem adequate to this situation, but a second, longer glance shows that the difficulty vanishes. In fact, since all the travel times are positive, we must have

$$t_{12} + t_{21} + f_1 > f_1 \qquad t_{13} + t_{31} + f_1 > f_1 \tag{8}$$

Consequently, neither $t_{12} + t_{21} + f_1$ nor $t_{13} + t_{31} + f_1$ can equal f_1, and so neither can yield the minimum in (7). It follows from (7) then, that

$$f_1 = \min\,(t_{12} + t_{24},\ t_{13} + t_{34}) \tag{9}$$

Substitution of this result back into (2) yields the two results

$$f_2 = \min\,[t_{21} + \min\,(t_{12} + t_{24},\ t_{13} + t_{34}),\ t_{24}]$$
$$= \min\,(t_{21} + t_{12} + t_{24},\ t_{21} + t_{13} + t_{34},\ t_{24})$$
$$= \min\,(t_{21} + t_{13} + t_{34},\ t_{24}) \tag{10}$$
$$f_3 = \min\,(t_{31} + t_{12} + t_{24},\ t_{34})$$

Unfortunately, these results, won after some considerable effort, merely state what was obvious in the beginning: the optimal route from 1 is either 1, 2, 4 or 1, 3, 4; the optimal route from 2 is either 2, 1, 3, 4 or 2, 4; and the optimal route from 3 is either 3, 1, 2, 4 or 3, 4. This conclusion may reinforce our confidence that Eq. (13.2) correctly and fully described the problem, but the foregoing elimination technique does not seem to be of any help in obtaining the values of the f_i if the number of points is large.

In the next chapter we shall introduce a simple yet powerful method for treating Eq. (13.2) which can be used to provide both an effective method of numerical calculation and a tool for ascertaining many properties of the solution.

Exercises

1. Solve the equation $x = \min\,(a_1 x + b_1,\ a_2 x + b_2)$ under the assumption that $b_1, b_2 > 0$, $0 < a_1, a_2 < 1$. (Hint: $x \le a_1 x + b_1$ implies that $x \le b_1/(1 - a_1)$ if $0 < a_1 < 1$).

2.   Write the Eqs. (13.2) for f_1, f_2, f_3 in Fig. 16 below if the destination is 4. By eliminating all other unknowns, express f_1, f_2, f_3 as minima of sums of the t_{ij}.

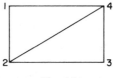

Fig. 16

3.   Suppose that there is a delay d_i in getting through the ith intersection. Show that the equations corresponding to the foregoing are

$$f_i = \min_{j \neq i} [t_{ij} + d_j + f_j].$$

15.   Why Not Enumeration Using Computers?

It is clear that as N increases the amount of effort required to determine the shortest route by the method of enumeration becomes prohibitive, at least by hand calculation. But what about computers operating at microsecond* speeds?

Consider a network of 12 points with the twelfth point as the destination and every pair of points joined by a link. Suppose that the starting point and the terminal point are fixed and we wish to enumerate all possible paths from the one to the other. Let us first of all determine the number of paths which pass through every vertex; although the total number of paths is much larger, this will give us the magnitude of the proposed enumeration. From the first point we can go to any of the 10 points which are neither the origin nor the terminal point, from that point to any of the 9 remaining points, and so on. There are thus

$$10 \times 9 \times 8 \cdots 2 = 10! \qquad (1)$$

of these paths. This number is a convenient measure of combinatorial problems, namely

$$10! = 3,628,800 \qquad (2)$$

This represents a considerable amount of hand calculation. Nevertheless, it is feasible to think in terms of 3,628,800 evaluations and comparisons with a computer that can, for example, perform an addition in a microsecond. Enumeration appears feasible.

Suppose, however, that there are 22 points. This is certainly not a large map. Then there are at least 20! different paths and clearly

* Recall that a microsecond is a millionth of a second.

$$20! > 10^{10}(10!) \tag{3}$$

To get some idea of the magnitude of this last number, let us start with the fact that the number of seconds in a year is a bit more than 3×10^7. To perform 20! operations at the rate of one a microsecond would thus require more than

$$\frac{10^{10}(10!)}{3 \times 10^{13}} \cong \frac{10^{10}(3 \times 10^6)}{(3 \times 10^{13})} = 10^3 \text{ years} \tag{4}$$

This figure will certainly be reduced as computers increase in speed.* Nonetheless, consider all of the improvements in computers that can be optimistically imagined, and then match them against the task of performing 100!, 1000! or 10,000! evaluations in a straightforward fashion.

It is clear that without the aid of mathematical techniques there is no hope of obtaining the desired numerical results. Brute force must be combined with bright force.

16. Graphs

Our discussion of logical trees and maps has now led us to a point where it is natural to introduce the general concept of a *graph*. By a graph we mean a figure consisting of a finite number of points in the plane (or in a three-dimensional space), and certain line segments joining pairs of points. For example, Fig. 17 shows a graph consisting of points A, B, C,

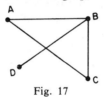

Fig. 17

D and segments AB, AC, BC, BD. We shall refer to the points as *vertices* or *nodes* and to the segments as *edges* or *links*.** The *theory of graphs* is the branch of mathematics which treats various abstract properties of such geometrical figures. We shall describe a few typical problems of graph theory in order to give the reader a notion of what kinds of questions arise.

First, however, we remark that graph theory as we are here describing it has nothing to do with the problems of "graphing" which the student

---

* And as parallelization of computer innards becomes commonplace.

** The terminology in the theory of graphs is at present not completely standarized. Note, by the way, that it may happen, as in Fig. 17, that two edges of a graph cross but the crossing point is not considered to be a vertex. In such a case, we could draw the graph in space in such a way that the edges do not intersect.

has no doubt encountered in his algebra, analytic geometry, or calculus courses. The problem in those courses was to represent a functional relation $y = f(x)$ by means of a graph, that is, a geometrical locus. In our present context, on the contrary, we do not think of a graph such as Fig. 17 as representing a functional relation at all. We wish to study the intrinsic properties of these graphs, just as in high school geometry we study intrinsic properties of triangles, circles, and so on, without thinking of any associated equations.

We have already had many examples of graphs. For example, each logic tree or tree diagram in Section 4 is a graph. Thus, Fig. 5 can be drawn as follows;

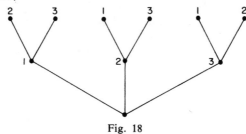

Fig. 18

The bottom vertex in this graph corresponds to the state of the urn before any ball is drawn. The three edges emanating from the bottom vertex correspond to the three possible ways in which the first draw may be made. The three vertices 1, 2, 3 correspond to the three possible outcomes of the first draw, and so on. In any event, Fig. 18 is in itself a graph. In this case, it was applied to the problem of drawing balls from an urn, but the same graph might be useful for some entirely different application.

The reader is probably familiar with some "combinatorial problems" from his earlier studies of mathematics. These problems arise typically in sampling and probability theory, and usually require a counting of the number of ways in which certain events can occur. As a simple example, let us consider our urn containing three numbered balls and ask for the number of drawings of two balls for which the number on the second ball drawn is larger than the number on the first ball drawn. It is obvious at once from the graph in Fig. 18 that there are three such drawings.

Many such counting or combinatorial problems are studied in graph theory. Here is a simple example. A graph is called *complete* if each pair

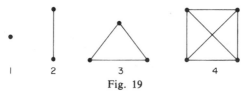

Fig. 19

of vertices is connected by an edge.

Figure 19 shows the complete graphs with 1, 2, 3, and 4 vertices, respectively. The number of edges in a complete graph with n vertices is easily seen to be 0, 1, 3, 6, 10 for $n = 1, 2, 3, 4, 5$, respectively. In general, the number of edges is the same as the number of ways of choosing 2 vertices out of n. As is well known, this number is given by the binomial coefficient $\binom{n}{2} = n(n-1)/2$.

It is also clear that any city street map or any road map is a graph, provided that we allow edges which are not necessarily straight. For example, the city map in Fig. 3 is a graph in which the edges represent street sections and the vertices represent street intersections. We realize now that *we have been using graphs in two separate ways in this chapter: on the one hand to represent actual street or road networks, and on the other hand to represent the logical tree of possible routes.*

We can now formulate the "shortest route problem" in the following very general way. Let there be a given graph with N vertices labelled 1, 2, $\ldots$, N. With each *directed edge* of the graph, let there be associated a positive number, denoted by t_{ij}, for the edge from vertex i to vertex j. For each initial vertex, consider all paths to the destination N.

The *shortest route problem* is the problem of finding among all these paths the path or paths for which the sum of the numbers t_{ij} is as small as possible.

We have used the term "directed edge" above because, as we have seen, t_{ij} and t_{ji} need not be equal. The notion of *directed edge* occurs frequently in graph theory, and we shall return to it again presently.

In our previous discussions, the numbers t_{ij} have represented times, the driving times for the various edges of the graph. Consequently, the path for which the sum of the numbers t_{ij} is least is the quickest path. On the other hand, we might let the t_{ij} be the lengths of the edges, and then the path for which the sum of the t_{ij} is least will be the shortest path. In other applications the numbers t_{ij} may have a different meaning. In all cases, however, it is customary to refer to the problem as the "shortest route problem."

17. The Problem of the Königsberg Bridges

It should be mentioned that the formulation of the shortest route problem involves two essential aspects. First, there is the graph itself. Second, there is the associated matrix* of numbers t_{ij}. In many of the

* The reader familiar with the general concept of a function will see that this associated matrix represents the values of a function defined on the set of all directed edges of the graph.

typical problems of graph theory, the second aspect is not present. As an example, let us consider the famous *problem of the Königsberg bridges,* which was solved by the mathematician Leonhard Euler in 1736. The city of Königsberg (now Kaliningrad) in East Prussia is located on the banks and on two islands of the river Pregel. The various sections of the city were joined by seven bridges, as indicated in Fig. 20. The problem is the following: Is it possible to make a walking tour of the city in such a way as to return to the starting point after having crossed each bridge once, and only once? Euler's solution of this problem was published in the Academy of Science in St. Petersburg (Leningrad), and is considered to be the origin of the theory of graphs.

Let us see how the problem can be turned into a problem on graphs. To do this, we represent each of the four sections 1, 2, 3, 4 of the city by a vertex, and each of the bridges by an edge connecting two vertices. The

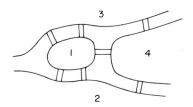

Fig. 20

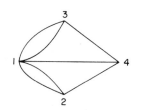

Fig. 21

result is shown in Fig. 21. The problem can then be stated in the follow- ing language. Is it possible to find a path in the graph which traverses each edge exactly once and returns to its starting point? As pointed out by Euler, it is quite easy to see that there is no such path in Fig. 21. The rea- son is that it would have to enter each vertex as many times at it leaves it; this is true even for the starting vertex. Hence there must be an even number of edges meeting at each vertex. Since this is not true for the graph in Fig. 21, no such path is possible.

Evidently it is possible to pose this same problem for any graph, as did Euler himself. A path in a graph which traverses each edge exactly once is now called an *Euler line* or *Euler circuit,* and a graph in which an

Euler line exists is called an *Euler graph*. We shall not pursue this topic at the present time, since our objective has been merely to give another example from the realm of graph theory. We note again that in this example there is no question of assigning lengths to the edges, nor of minimizing path lengths. The question is one of the *existence* of a path of a specified kind. This is a problem of topological type in which distances play no role.

However, the methods we use in what follows enable us to treat the problem of determining a circuit with the minimum number of repeated links. This, in turn, allows us to treat questions of the foregoing type by quantitative means. See the Miscellaneous Exercises at the end of Chapter Eight.

A few more examples from graph theory appear in the Exercises below, and many more applications will be encountered in the rest of the book.

Exercises

These exercises are designed to introduce the reader to additional problems and concepts of the theory of graphs. However, knowledge of these concepts will not be assumed in the rest of this book. These pages may be safely skipped in a first reading.

1. The edges of a graph which have a vertex A as an endpoint are said to be *incident* to A. The number of such edges is called the *local degree* at A. What is the local degree of each vertex in a complete graph with n vertices?

2. Show that in any graph the sum of the local degrees of all vertices is an even number. Deduce that any graph has an even number of vertices which have odd degree.

3. (a) Is it possible to construct a graph having five vertices of respective local degrees 1, 2, 3, 4, and 5?
   (b) Construct a graph with four vertices, each of degree 2.
   (c) Construct a graph with four vertices, two having degree 2 and two having degree 3.
   (See the book of Berge cited below for general theorems on the existence of graphs having preassigned local degrees.)

4. An *arc* is a route in a graph that goes through no vertex more than once. For example, in Fig. 21 the route [1, 2, 4] is an arc but [1, 2, 4, 1, 3] is not. A graph is called *connected* if each pair of its vertices can be connected by an arc. With these concepts in hand, we can now precisely define a *tree* as a connected graph that contains no circuits.

Check Figs. 6, 18, etc., to see that they fit this definition of tree. A tree with *n* vertices has how many edges?

5. Given *n* cities, how shall a network of roads of minimum total length be constructed so that it is possible to travel from any one city to any other? Roads may cross each other outside the cities. These crossings can be called junction points and there may be as many as necessary to minimize the total length. The roads are assumed to be straight line segments between cities and junction points. Solve this problem for *n* = 3. In graph theoretical language the problem can be stated as follows: given *n* points in the plane, find the shortest tree whose vertices include these *n* points. This is often called "Steiner's problem," and is quite difficult. See

H. S. M. Coxeter, *Introduction to Geometry*, Wiley, New York, 1961, 21.

R. Courant and H. Robbins, *What is Mathematics?* Oxford Univ. Press, New York, 1941.

A. F. Veinott, Jr. and H. M. Wagner, "Optimal Capacity Scheduling," RM-3021-PR, 1962, The RAND Corp.

J. M. Hammersley, "On Steiner's Network Problem," *Mathematika* **8** (1961) 131-132.

6. A variation of Steiner's problem occurs in printed circuit technology. Suppose that *n* electrical junctions are to be connected with the shortest possible length of wire and moreover that the wires must run in the horizontal and vertical directions. Solve this problem for *n* = 3 and *n* = 4. See

M. Hanan, "On Steiner's Problem with Rectilinear Distance," *SIAM J. Appl. Math.* **14** (1966) 255-265.

E. N. Gilbert and H. O. Pollak, "Steiner Minimal Trees," *SIAM J. Appl. Math.* **16** (1968) 1-29.

For an interesting account of the application of these ideas to the theory of evolution, see

A. W. F. Edwards and L. L. Cavalli-Sforza, *Reconstruction of Evolutionary Trees*, Systematics Association Publication No. 6, London, 1964.

7. Two vertices of a graph are *adjacent* if they are end points of the same edge. A graph is called *2-chromatic* if its vertices can be painted with 2 colors in such a way that no two adjacent vertices have the same color. It is *3-chromatic* if its vertices can be painted with 3 colors in such a way that no two adjacent vertices have the same color, and so on. The smallest number γ for which the graph is γ-chromatic is called the *chromatic number* of the graph.

(a)   Find the chromatic number of each of the graphs in Fig. 19.
(b)   Find the chromatic number of the graph in Fig. 21.
(c)   What can be said about the chromatic number of a tree?

8. In a certain town, the blocks are rectangular, with the streets (of zero width) running East-West, the avenues North-South. A man wishes to go from one corner to another *m* blocks East and *n* blocks North. The shortest path can be achieved in many ways. How many? What is the relation between this problem and Pascal's triangle?

    Hint: Let $f(m, n)$ denote the number of ways in which the shortest path can be achieved. Show that f satisfies the functional equation (recurrence relation)

    $$f(m, n) = f(m - 1, n) + f(m, n - 1)$$

See

University of Wisconsin Mathematical Talent Search., *Amer. Math. Monthly* 73 (1966) 401.

Research Problem: Discuss the relation between the problem of coloring a map and the theory of graphs (see the books of Ore and Berge cited at the end of the chapter and the Miscellaneous Exercises at the end of Chapter Eight.)

18. Systems and States

In many physical applications, it is customary to refer to the *state* of a *system*. For example, in Newtonian mechanics one is often interested in the "system" consisting of a mass particle (or some physical object) moving in a certain environment. Usually one refers to the position and velocity of the particle at a specified time as the "state" of the system at that time. Knowing the physical laws governing the motion of such a particle and its state at a certain time, one can predict the state at other times. That is, the physical laws determine how the system passes from one state on to subsequent states.

This terminology can profitably be carried over to the shortest route problem. Think once again of a car and driver moving along certain streets. Let us agree to refer to the city streets, the car, and the driver as the "system" under consideration. The "state" of this system at a specified time could be defined to be the location of the car at that time. The shortest route problem could then be stated as the problem of finding the least time in which the system can pass from the initial state (in which the car is at the starting vertex) to the final state (in which the car is at the terminal vertex).

In Newtonian mechanics there is ordinarily an infinity or "continuum" of states to be considered. For example, a planet moving around the sun is thought of as moving in a continuous way from state to state. It can occupy infinitely many locations in the course of its journey.

In our work, however, we need distinguish only a finite number of states. In fact, we have seen that we only need to consider the times t_{ij} required for a car to go from one vertex to another. Thus, we could define the possible states for a map with N vertices as follows:

State Number	Description of System
S1	The car is at vertex 1.
S2	The car is at vertex 2.
...	...
SN	The car is at vertex N.

Of course when we define the states in this way, we are making an abstract *model* of the real situation, and in so doing we are ignoring many things which we regard as of minor importance in relation to our objective. For example, none of the variations of speed or other events which occur *between* two vertices are taken into account. This is the way of science: to focus attention initially on an idealized situation in which relations can be clearly discerned. Then, step-by-step, we introduce more realistic features.

It is easy to think of circumstances in which a different model is required to encompass some aspects of reality that we have thus far ignored —such as a penalty (in time) for making a right or left turn at an intersection rather than going straight through. Consider the map shown in Fig. 22, with vertex 1 the starting point and vertex 5 the destination. The times required to go from one vertex to another are shown on the connecting edges. According to our previous model, this system is one with five states, which we may now denote by $S1, S2, S3, S4, S5$, corresponding to the car's being at vertex 1, 2, 3, 4, 5 respectively. There are 4 routes from $S1$ to $S5$, the shortest being $S1-S3-S4-S5$.

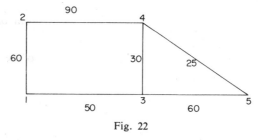

Fig. 22

Suppose, however, that there is an additional penalty of 5 time units whenever a car makes a left or right turn at an intersection. Then the time for going from vertex 1 to 3 to 4 is not $50 + 30$ but rather $50 + 5 + 30$;

the time for going from 1 to 3 to 5 is still 50 + 60. Before we can deal with this problem by the methods we have thus far used, we must construct a new model. We can, for example, think of the state as denoting the link on which the car is situated just prior to entering an intersection. To be specific, let us label the directed links 1, 2, 3, 4, 5, 6, 7 as shown in Fig. 23. Note that the link joining vertices 3 and 4 must be represented by two directed links. The other links are represented by one-way directed edges, since the opposite direction can obviously not be optimal in our particular problem.

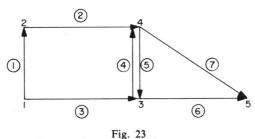

Fig. 23

In this new version, there are nine states, as listed below:

State Number	Description of System
S0	The car is at the starting vertex.
S1	The car has just traversed arc 1 and is about to enter vertex 2.
...	...
S4	The car has just traversed arc 4 (up) and is about to enter vertex 4.
S5	The car has just traversed arc 5 (down) and is about to enter vertex 3.
...	...
S8	The car is at the terminal vertex.

It is necessary to extend our definition to include a description of the way the transition time from one state to another is determined. We define the time to go from state Si to an allowable state Sj to be the time required for a turn (if any) plus the time for traversing the subsequent link up to its end vertex. No turn penalty will be counted at the terminal point. For example, the transition from $S0$ to $S1$ takes 60 seconds, that from $S1$ to $S2$ takes 5 + 90 = 95 seconds, that from $S2$ to $S5$ takes 5 + 30 = 35 seconds, and that from $S2$ to $S4$ is impossible. For convenience we say that there is no time connected with going straight through an intersection. The matrix of transition times is shown in Table 12. We find that the

TABLE 12

To From	S0	S1	S2	S3	S4	S5	S6	S7	S8
S0	0	60		50					
S1		0	95						
S2			0			35		30	
S3				0	35		60		
S4					0			30	
S5						0	65		
S6							0		0
S7								0	0
S8									0

optimal route is now $S0$–$S3$–$S6$–$S8$, corresponding to the path from vertex 1 to vertex 3 to vertex 5.

Observe that the second model for the map with 5 vertices has resulted in a system with 9 states. If we wish, we can make a graph showing these 9 states and the possible connections between them (see Fig. 24). In this way, it is still possible to conceive of the states as represented by vertices, not of the original map, but of a new *state graph*.

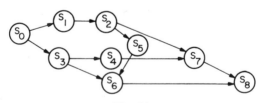

Fig. 24

In going from our first model to our second model we have obtained a more realistic description. At the same time, however, we have increased the number of states and thereby lengthened the time required to compute a solution of the problem. As the number of vertices in the original map is increased, it becomes important to decide whether the gain in realism makes the greater computational effort worthwhile. Questions of this nature require both skill and experience to answer. The proper balancing of realism versus effort is a fine art, which requires the usual grounding in experience, as well as a certain inborn ability.

Exercises

1. List all the states for the map in Fig. 3 (with the revised definition of the states), if vertex 8 is the destination.

2. Consider the map with 8 vertices shown below. Assume that each

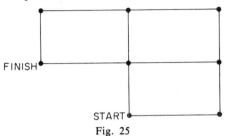

Fig. 25

link can be traversed in one unit of time, and that there is a penalty of one unit for each right turn and of 10 units for each left turn (left turns are sometimes illegal or impossible). Show that the optimal path goes through one vertex twice, but that no state (with the revised definition of state) is encountered twice.

3. Obtain an analytic formulation of the problem of determining an optimal route when the time to traverse the link connecting i and j depends upon the point one was at before coming to i. Let $t(k, i, j)$ denote the time required to traverse the link ij, having just been at k. (Note: The dependence on the present and immediate past state is analogous to the situation in mechanics where both position and velocity are required to determine the subsequent behavior.)

19. Critical Path Analysis

A problem which is akin to the shortest route problem, and of great importance in industrial applications, is the problem of the determination of a "critical path." As an example, let us consider the scheduling of the operations which are necessary in the construction of a house. Among these we may mention:

    (a)   Purchase of site,
    (b)   Obtaining a building permit,
    (c)   Clearing and excavation,
    (d)   Pouring of foundations,

and so forth. Some operations can be begun only if certain others are already completed, but some operations can be carried on simultaneously. One can ask, then, for the least time in which the whole job can be finished.

Moreover, the contractor will want to know the anticipated completion time for each operation, in order to be able to schedule the arrival dates for the various subcontractors and materials.

For the sake of a simple illustration let us suppose that there are seven jobs or operations in all, designated $J1, J2, \cdots, J7$, and consider the graph in Fig. 26. The seven numbered circles represent the basic operations. A directed segment and the number beside it indicate that the operation at the end of the arrow cannot begin until this number of weeks after the commencement of the operation at the beginning of the arrow. Thus, $J2$ cannot begin until 4 weeks after $J1$ begins, $J3$ cannot begin until 7 weeks after $J1$ begins and 2 weeks after $J2$ begins, and so on. We assume that a job is completed when all subsequent jobs have begun. For example, $J1$ is completed in 7 weeks, $J2$ is completed 3 weeks after it begins, etc.

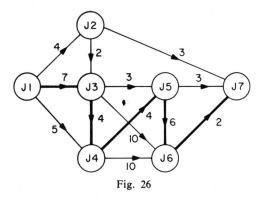

Fig. 26

It is important to observe that on a graph of this type all edges are directed, and between any two vertices there cannot be an edge in both directions. In fact, there can be no closed circuit in the graph, since a circuit such as $J1 \rightarrow J2 \rightarrow \cdots J1$ would indicate that $J1$ cannot begin until a number of weeks after $J1$ begins, which is nonsensical.

Now if all jobs are begun at the earliest possible moment, and completed in the specified time, what will be the time required to complete the whole job? It is the time for the path from $J1$ to $J7$ of the *greatest* possible total time, or the slowest path, since no job can be begun until the most time-consuming of its predecessors has been completed. In our example, this maximal path is $J1, J3, J4, J5, J6, J7$, shown in Fig. 26 in bold face.

The path of largest time from start to completion is called the *critical path*, since any delay in going from one job to the next on this path will result in a delay in completing the whole job. On noncritical paths, such delays need not increase the time for the whole job. For example, in Fig. 26 if the step from $J1$ to $J3$ takes 9 weeks instead of 7, the least time to complete the whole project will increase from 23 to 25 weeks. However, if the step from *J*2 to *J*7 takes 5 weeks instead of 3, there will be no increase in the duration of the project since this step can still be completed in 9 weeks total, well before the 23 weeks for the project.

It is not hard to give a method for the determination of the slowest path which is similar to the one we have discussed for the quickest path. Let us define

$$t_{ij} = \text{time associated with the directed link from}$$
$$\text{vertex } i \text{ to vertex } j \tag{1}$$

and let

$$f_i = \text{time for the longest path from vertex 1 to vertex } i \tag{2}$$

for each $i = 1, 2, \cdots, N$, where N is the terminal vertex. Clearly $f_1 = 0$. We leave it to the reader to derive the basic equations

$$f_j = \max(t_{ij} + f_i) \qquad j = 1, \cdots, N, \tag{3}$$

where the maximum is over all vertices i for which there is a directed edge from i to j.

For Fig. 26, these equations are as follows:

$$
\begin{aligned}
f_1 &= 0 \\
f_2 &= 4 + f_1 \\
f_3 &= \max(7 + f_1, \ 2 + f_2) \\
f_4 &= \max(5 + f_1, \ 4 + f_3) \\
f_5 &= \max(3 + f_3, \ 4 + f_4) \\
f_6 &= \max(10 + f_3, \ 10 + f_4, \ 6 + f_5) \\
f_7 &= \max(3 + f_2, \ 3 + f_5, \ 2 + f_6)
\end{aligned}
\tag{4}
$$

The values of all the unknowns can be found successively from these equations, since in each equation only previously computed f's are needed. Thus we get $f_2 = 4$, $f_3 = 7$, $f_4 = 11$, $f_5 = 15$, $f_6 = 21$, $f_7 = 23$, and it is also easy to deduce the critical path itself.

Since the graph of a scheduling problem of this kind contains no circuits, the solution of the equations (1) can always be performed one-by-one as in this example. Thus, the Eqs. (3) are intrinsically simpler in this case than were the Eqs. (13.1) for the shortest route problem. In particular, it is unnecessary to use the method of successive approximations as explained in Chapter Two.

We have assumed that the values assigned to the edges of our graph are precisely known. In practice, these data are far from exact. A method called PERT (Program Evaluation Research Task) has been developed by the Special Projects Office of the U. S. Navy, and others subsequently, to solve scheduling problems of this sort under more realistic assumptions. In this method, the vertices of the graph represent events and the edges represent the intervening operations. The times required for these operations are supplied by specialists, who estimate the values on the basis of an assumed probability distribution. The total mean delay for the critical path is then computed and the total cost of the operations evaluated. Once this is done, consideration is given to possible modifications of the operational steps in the process which might achieve the same goal. Such a modification yields a modified graph, hence possibly reduced time or cost for completion of the project. More detailed discussions of the PERT method can be found in the references cited at the end of the chapter.

Still other scheduling problems are discussed below in Chapter Eight.

Exercises

1.   For the example in Fig. 26, what is the maximum delay in the step from $J1$ to $J4$ which will not delay the completion of the whole project? Answer the same question for the step from $J3$ to $J6$.

2.   For the map of Fig. 3, § 3, find the slowest path from vertex 0 to vertex 8.

3.   Give a proof for Eqs. (19.3).

20.   Summing Up

Let us briefly review the principal features of this chapter. Our first objective was to introduce a class of problems of intrinsic mathematical interest and, in addition, of widespread application in the economic and engineering domains which require methods quite distinct either from those of elementary algebra or calculus.

In so doing we encountered matrices, trees, and graphs and were led to an abstract and general version of our problem as "Shortest route problem on an arbitrary graph." Furthermore, we introduced a new class of equations and in the consideration of equations of this nature we were forced to examine and analyze some of the differences between a theoretical solution, in this case one obtained by direct enumeration of possibilities, and a feasible solution. The criterion here is time, the time required to obtain the solution. The intellectual content of the original puzzle, that of finding the quickest route between home and the office, has thus been amply demonstrated. It remains, however, to demonstrate that we do possess an effective method for determining paths of minimum time based upon the equations for the quantities f_i. Chapter Two will be devoted to this aim. Otherwise, we will have succeeded only in transforming an unsolved problem of one kind into an unsolved problem of a different kind. There is often considerable merit to this since there is always the possibility that there is someone who already knows the solution of the new "unsolved" problem. But it is not a completely satisfying endeavor.

Bibliography and Comments

The approach to the routing problem appearing here was first given in

B. Bellman, "A Routing Problem," *Quart. Appl. Math.* **16** (1958) 87–90.

A detailed account of a number of other methods to treat problems of this nature, together with comparisons and evaluations, will be found in

S. Dreyfus, *An Appraisal of Some Shortest Path Algorithms*, RAND Corporation, RM-5433-PR, Oct. 1967, Santa Monica, California.

Furthermore, a great deal of work has been devoted to the determination of second and third best paths, and so on; see

M. Pollack, "Solutions of the *k*th Best Route through a Network—A Review," *J. Math. Anal. Appl.* **3** (1961) 547–559.

The approach used in this chapter, and in the following chapters, is an application of the theory of dynamic programming, a modern mathematical theory devoted to the study of multistage decision processes, and those processes which can profitably be interpreted in this fashion. In particular, the theory has extensive applications to design and control processes in the fields of economics and engineering, to conceptual and control processes in psychology and the biomedical area, and furnishes new insight into many classical areas of analysis such as the calculus of variations.

Elementary expositions may be found in

R. Bellman, *Dynamic Programming*, Princeton Univ. Press, Princeton, New Jersey, 1957.

R. Bellman and S. Dreyfus, *Applied Dynamic Programming*, Princeton Univ. Press, Princeton, New Jersey, 1959.

S. Dreyfus, *Dynamic Programming and the Calculus of Variations*, Academic Press, New York, 1965.

A. Kaufmann, *Graphs, Dynamic Programming, and Finite Games*, Academic Press, New York, 1967.

A. Kaufmann and R. Cruon, *Dynamic Programming: Sequential Scientific Management*, Academic Press, New York, 1967.

A. Kaufmann and R. Faure, *Introduction to Operations Research*, Academic Press, New York, 1967.

The books by Kaufmann also contain excellent introductory discussions of graph theory. Further material of importance may be found in

O. Ore, *Graphs and Their Uses*, Random House, New Mathematical Library, New York.

O. Ore, *Theory of Graphs*, American Mathematical Society Colloquium Publications, **38**, 1962.

C. Berge, *The Theory of Graphs*, Wiley, New York, 1962.

The continuous version of the routing problem is the determination of geodesics on surfaces, a problem of considerable analytic, geometric, and physical interest. See

L. A. Lyusternik, *Shortest Paths; Variational Problems*, Macmillan, New York, 1964.

The book by Dreyfus also considers some stochastic versions of the routing problem. Two interesting survey papers on graph theory are

F. Harary, "Unsolved Problems in the Enumeration of Graphs," *Publ. Math. Inst. Hungarian Acad. Sci.* **5** (1960) 63–94.

F. Harary, "Some Historical and Intuitive Aspects of Graph Theory," *SIAM Rev.* **2** (1960) 123–131.

For an interesting way in which shortest routes occur in physiology, see

A. Munck, "Symbolic Representation of Metabolic Systems," *Mathematical Biosciences* (forthcoming).

§ 13. See

F. T. Boesch, "Properties of the Distance Matrix of a Tree," *Quart. Appl. Math.* **26** (1969) 607–610.

§ 17. See

L. Euler, The Seven Bridges of Königsberg, *World of Mathematics*, Vol. I, p. 573, Simon and Schuster, New York 1956.

S. Kravitz, "Solving Maze Puzzles," *Math. Mag.* **38** (1965) 213–216.

§ 18. See

J. F. Shapiro, "Shortest Route Methods for Finite State Space, Deterministic Dynamic Programming Problems," Technical Report 31 (1967), Operations Research Center, Mass. Inst. Technol., Cambridge, Mass.

§ 19. For a discussion of critical path analysis and the PERT method, see

D. R. Fulkerson, "Expected Critical Path Lengths in PERT Networks," *Oper. Research* **10** (1962) 808–817.

D. G. Malcolm, J. H. Roseboom, C. E. Clark, and W. Fazar, "Application of a Technique for Research and Development Program Evaluation," *Oper. Research* 7 (1959) 646–669.

M. Montalbano, "High-speed Calculation of the Critical Paths of Large Networks," *IBM Systems J.* **6** (1967) 163–191.

An interesting feature of the last paper is the inclusion of a description in Iverson's language of the algorithm used.

Chapter Two

THE METHOD OF SUCCESSIVE APPROXIMATIONS

1. Introduction

In the previous chapter, we were led to the equations

$$f_i = \min_{j \neq i} [t_{ij} + f_j], \qquad i = 1, 2, \cdots, N - 1,$$
$$f_N = 0 \tag{1}$$

which are satisfied by the minimal times associated with a map containing N intersection points, or, as is more convenient to say, N vertices. This is a set of $N - 1$ simultaneous nonlinear equations for the desired quantities f_i. Consequently, it is not at all clear that these equations can be used in any meaningful way to obtain a solution to the original problem, which is to say, to determine both the values f_i, $i = 1, 2, \cdots, N - 1$, and the route followed in traversing the minimal path. We possess a large number of systematic methods for solving systems of linear equations, but this is not the case for nonlinear equations. Each set is an individual challenge.

Unless, therefore, we can exhibit a feasible technique for wresting numerical solutions from (1), given numerical values for the t_{ij}, we will be forced to admit that these relations are decorative, but hardly functional, and that the original problem remains open.

Fortunately, a systematic technique does exist; as a matter of fact, many excellent techniques are available for the solution of the original problem. References to some of these will be found at the end of the chapter. In presenting the technique that we apply we are guided by the following pragmatic considerations:

(a)   The method is to be simple in concept.

(b)   The method is to be simple to apply and well suited to the capabilities of the contemporary digital computer.

(c)   It must be applicable to a wide range of equations other than (1).

(d)   It must be a method which can be established rigorously in the
present context without undue effort, or demands on mathematical training.

(e)   The detailed applications of the method are to be motivated by
the intrinsic nature of the problem under consideration.

It is quite remarkable that a method with all of these sterling qualities
exists. However, it does indeed and bears the name of the "method of
successive approximations." It is one of the keystones of mathematical
physics and applied mathematics in general, and should be a part of the
toolchest of every mathematician.

This chapter is devoted to an explanation of the method of successive
approximations as it applies to the equations of (1), and to other equations
of perhaps more familiar type. In this connection we shall also discuss and
illustrate the general concept of an algorithm and construct algorithms for
computer use for the particular problem under consideration. Additional
computational devices for large scale problems of this nature and examples
will be given in Chapter Three.

At this point it is important to point out that although we consider
the foregoing equations to be fundamental and have focused our attention
on them, we have not yet validated this emphasis. It is essential at some
stage to show that they completely and unambiguously determine the solution to the original verbal problem. This means that we must establish
existence and uniqueness of the solution, and also show that all minimal
paths are determined in this fashion. The first part will be done in Chapter
Four, the second part here.

2.   A Simple Example of Successive Approximations

In order to illustrate the basic idea of the method we shall employ, let
us temporarily sidetrack and consider a more familiar type of problem
first. Let us take the problem of solving the quadratic equation

$$x + 1 = x^2 \tag{1}$$

As we know from algebra, the positive root of this equation is given by
the expression

$$x = \frac{1 + \sqrt{5}}{2} = 1.61803 + \tag{2}$$

However, if the equation were of the form $x + 1 = x^7$, we would no longer possess any simple explicit algebraic representation of the solution which could be used to provide a numerical solution. Let us then see if we can find a method of obtaining an approximate solution of (1) which can be used as well for the numerical study of more general equations, such as the seventh degree equation just mentioned.

Since $1 + 1 > 1^2$, $2 + 1 < 2^2$, we see that the positive solution of (1) lies between 1 and 2, as indeed (2) verifies.

Let us take $x = 1$ as an initial guess; call this value x_1. We now determine a second guess, which we call x_2, by using the approximate value in the left side of equation (1), thus:

$$x_1 + 1 = 1 + 1 = x_2^2 \qquad (3)$$

Since $x_1 = 1$, we have $x_2 = \sqrt{2}$. A third guess is obtained by repeating this process using the new approximate value in the left-hand side of (1)

$$x_2 + 1 = \sqrt{2} + 1 = x_3^2 \qquad (4)$$

whence $x_3 = \sqrt{1 + \sqrt{2}} = 1.554 -$. Continuing in this fashion, we see that the nth approximation, x_n, is obtained from the $(n-1)$st, x_{n-1}, by solving the equation

$$x_{n-1} + 1 = x_n^2 \qquad (5)$$

From this we see that $x_n = \sqrt{x_{n-1} + 1}$. Starting with $x_1 = 1$, we obtain in this way the values $x_2 = 1.414$, $x_3 = 1.554 -$, $x_4 = 1.598 +$, reasonable approximations to the value in (2).

It can easily be shown that the successive values obtained by this procedure are increasing, that is to say that

$$x_1 < x_2 < x_3 < \cdots < x_{n-1} < x_n < \cdots . \qquad (6)$$

We shall prove this by the method of mathematical induction. Let us see how this goes. It is certainly true that $x_1 < x_2$ since $x_1 = 1$ and $x_2 = \sqrt{2}$. Now we assume as the inductive hypothesis that $x_{n-1} < x_n$ and try to prove that $x_n < x_{n+1}$. Since $x_{n+1} = \sqrt{1 + x_n}$, what we need to prove is that $x_n^2 < 1 + x_n$. But since $x_n^2 = 1 + x_{n-1}$, this reduces to the inequality $x_{n-1} < x_n$, which was assumed as hypothesis. Consequently the inductive hypothesis implies that $x_n < x_{n+1}$, as we sought to prove.

It is also readily proved inductively that each value x_n is less than x, the positive root of (1) given by (2). For certainly $x_1 < x$, and the hypothesis $x_{n-1} < x$ implies

$$x_n = \sqrt{x_{n-1} + 1} < \sqrt{x + 1} = x \qquad (7)$$

since x satisfies (1).

The fundamental convergence theorem for an increasing sequence of real numbers with an upper bound tells us that the sequence of values x_1,

$x_2, \cdots, x_n \cdots$, has a limit.* Call this limit y. Then from (5), we see that

$$1 + y = y^2 \tag{8}$$

Since (1) has only *one* positive root, we know that $y = x$. Thus, we know that our sequence of approximations converges to the desired solution.

Let us note parenthetically that we have established the existence of a feasible algorithm for solving (1). However, there still remain a number of interesting and important lines of investigation concerned with the determination of accurate, rapid, and simple types of successive approximation. Some of this will be discussed in Chapter Three.

The reader unfamiliar with the notion of limit may feel out of his depth at this point. Let him be assured that no further reference to this nonalgebraic concept will be made in what follows as far as the treatment of Eq. (1.1) is concerned. The reason for this is that the method of successive approximations we employ below to treat the equation in (1.1) yields the solution in a finite number of steps. Hence, the discussion remains on an elementary level. This point is discussed in § 14. However, in § 7 we shall employ a modicum of calculus to obtain the tangent to a point on a parabola in connection with an illustration of the use of successive approximations.

If we start with a different initial approximation, say $z_1 = 2$, we have $z_2 = \sqrt{z_1 + 1} = \sqrt{5}$, and, generally, as before,

$$z_n = \sqrt{z_{n-1} + 1} \tag{9}$$

The numerical values of the successive approximants are, of course, now quite different, namely $z_1 = 2$, $z_2 = 1.732+$, $z_3 = 1.653-$, $z_4 = 1.629-$. Since $z_1 > x$, we obtain in this way a sequence $\{z_n\}$ with the property that

$$z_1 > z_2 > \cdots > z_n > \cdots > x \tag{10}$$

We leave it to the reader to establish this inductively.

It follows that we can obtain arbitrarily accurate upper and lower bounds for the desired root. This is an important advantage when we can manage it, which is unfortunately not very often.

Exercises

1. How many steps do we have to go to obtain $\sqrt{2}$ accurately to five significant figures; to six, seven?

2. Obtain in this way the positive root of $x + 1 = x^3$; $x + 1 = x^7$; to

---

* At this point, we feel compelled to insert a few paragraphs concerning the limit of x_n as n increases without bound. The reader unfamiliar with the notion of limit may safely omit the next two paragraphs.

five significant figures in each case.

3. Obtain the positive root of $x + 3 = 3^x$ to three significant figures in this way.

4. What advantage do we obtain by averaging the upper and lower bounds?

3. A Simple Example of the Failure of the Method

Suppose that we try the method in the following fashion. Write the equation in the form

$$x = x^2 - 1 \tag{1}$$

Take $x_0 = 2,$* and compute the successive approximations

$$x_1 = x_0^2 - 1 = 3$$
$$x_2 = x_1^2 - 1 = 8$$
$$x_3 = x_2^2 - 1 = 63 \tag{2}$$
$$\vdots$$

It is apparent that the sequence of values $x_0, x_1, x_2, \cdots$ does not yield more and more accurate estimates of the positive root of (1). As a matter of fact, we get further and further from the solution at each stage.

If, on the other hand, we take $x_0 = 1$, we encounter the following amusing situation:

$$x_1 = x_0^2 - 1 = 0$$
$$x_2 = x_1^2 - 1 = -1$$
$$x_3 = x_2^2 - 1 = 0 \tag{3}$$
$$\vdots$$

Once again, it is obvious that the sequence of approximations is not zeroing in on the root of (1).

We see then that the method of successive approximations is powerful, but not infallible. This quality of algorithms is a considerable part of what makes mathematics interesting. We possess many techniques which work well when applied adroitly, but not necessarily otherwise. There is thus always the element of art in the specific utilization of a general procedure. Pertinent to this is the remark of the famous mathematician A. Hurwitz, "It is easier to generalize than particularize."

* We shall alternate between using the subscripts 0 and 1 to denote an initial guess to familiarize the reader with both notations. There are, as usual, advantages and disadvantages to both.

Observe also that one of the advantages of a fast computer lies in exhibiting rapidly the utility or fallibility of a particular algorithm. This is a most important point. If we can try several alternative methods one after the other, or simultaneously,* we have an improved chance of obtaining a solution.

As far as Eq. (1.1) is concerned, we are fortunate in possessing a guaranteed method, as we shall see.

Exercises

1. Write the equation in the form $x = (1 + x)/x$, and take $x_0 = 1$. Then write $x_1 = 1 + 1/x_0$, $x_2 = 1 + 1/x_1$, $x_3 = 1 + 1/x_2$, $\cdots$. Calculate the first six values and compare their values with the positive root of $x^2 = x + 1$.

2. Show that the approximations in Exercise 1 satisfy $x_0 < x_2 < x_4 < \cdots$, and that $x_1 > x_3 > x_5 > \cdots$.

3. Show that $x_0 < x_2 < x_4 < \cdots < x < \cdots x_5 < x_3 < x_1$, where x is the positive root of (1) above. A situation of this sort is very desirable in numerical work since it enables us to stop the calculation whenever the upper and lower approximating values are sufficiently close to each other.

4. Algorithms

One of the principal points to be noted in what has preceded is that the method of successive approximations constitutes a so-called algorithm or algorithmic process for solving equations of a certain class in terms of a succession of elementary arithmetic operations. The concept of an algorithm is one of the basic concepts of mathematics and therefore it is worth dwelling upon. By an algorithm we mean a systematic procedure for producing the answer to any problem of a given kind in terms of a specified set of operations. Or, we may say that an algorithm is a list of instructions for producing this answer.

The critical reader may well object that this definition is vague and imprecise, since there is some question as to what is meant by "systematic procedure," "list of instructions," "specified set of operations," or "any problem of a given kind." Well aware of this, in recent years, mathematicians and logicians have attempted to form precise definitions, and have thus created what is called that "theory of algorithms" and the theory of "computability." We shall not concern ourselves at this stage with these

---

* We are thinking here of the important new development of parallelization.

abstract theories, but rather shall attempt to create a working definition by giving many examples of algorithms.

A simple example of an algorithm is the set of rules learned in elementary school for adding two numbers given in decimal form. Such rules might be expressed in the following way:

(1) Add the rightmost digits. As an example suppose 54 and 78 are to be added. Then in this step we add 4 and 8 to get 12.

(2) If the sum is less than or equal to 9, record it as the appropriate digit in the answer. If it exceeds 9, record the right digit of the sum as the digit of the answer in this column, and carry a one to the next column to the left. In our example, we could perform this step by writing

$$
\begin{array}{r}
1 \\
54 \\
78 \\
\hline
2
\end{array}
$$

(3) Add the digits in the next column and add the carry if any. Repeat step (2). In our example, the result could be written

$$
\begin{array}{r}
11 \\
54 \\
78 \\
\hline
32
\end{array}
$$

(4) Repeat step (3) as long as possible, stopping when there are no entries to be added in a column and no carry into this column. In our example, the last step would be to record the 1 from the last carry, giving 132 as the final answer.

When the rules are written down in this way, one sees that they are fairly complicated, and involve various conditional rules of the form "if such and such, do so and so, but otherwise do something else." This is typical of many algorithms, and the ability of a modern computing machine to easily handle branching situations of this nature is a great advantage. Of course, the student in elementary school probably never sees a set of written rules such as the above. Instead, he learns to go through all of the necessary steps by doing a very large number of drill problems. Nevertheless, the set of rules, or as we shall say the algorithm, does exist and is used implicitly.

Algorithms need not be of an arithmetic kind. For example, consider the problem of sorting into alphabetical order four cards to be read one after another, each card containing a single letter of the English alphabet. An algorithmic process to solve this problem is as follows.*

---

* This is a simplified version of what is often called *bubble sorting*. See the Exercises for more general and more efficient versions.

(1) Read the letter on the first card.
(2) Read the letter on the next card. If it should precede the letter on the previous card, interchange these two cards. Otherwise, go on to the next step.
(3) Repeat step (2) until there are no more cards.
(4) Repeat steps (1) to (3).
(5) Repeat steps (1) to (3) again.

For example, suppose the cards originally contain the letters D, C, B, A, in this order. In step (1) and the first application of step (2) we compare the first two cards and interchange them, giving the order C, D, B, A. Next, step (2) is repeated and the second and third cards are interchanged, giving C, B, D, A. Another repetition of step (2) gives C, B, A, D. Thus, one turn through steps (1), (2), (3) moves the D to the end. In step (4), we repeat this process, after which the order is B, A, C, D. Finally, in step (5) we repeat the process again, obtaining at length the correct alphabetical order A, B, C, D.

Of course, such a simple job as this could be done almost in one glance by a secretary. However, if the job is very large, for example to put in alphabetical order all the names that eventually constitute the New York City telephone directory, then a precisely formulated algorithm and the assistance of computing machines are important. Once again, we face a choice of possible algorithms graded according to speed, probability of error, ability to be carried out by different levels of personnel, etc.

We have so far discussed examples of the ability of the computer to carry out elementary arithmetic operations and operations involving order comparisons. What about the ability to calculate $\sqrt{2}$ or log 2? Since we can only store rational numbers, and perform arithmetic operations which yield other rational numbers, we know in advance that we can never calculate $\sqrt{2}$ exactly using a digital computer.

The fact that $\sqrt{2}$ has no representation of the form p/q where p and q are integers was wellknown to the Greeks and we urge the reader to carry through a demonstration for himself based upon a *reductio ad absurdum*. That log 2 has no such representation is considerably more difficult to establish.

These are interesting results and well worth knowing, but of no particular consequence as far as the numerical solution of problems is concerned. What is desired in practice is a set of algorithms for obtaining arbitrarily accurate approximations. In particular, we want algorithms which obtain these results rapidly and reliably. In more advanced mathematics, approximations are far more important than exact results, and this holds true generally throughout science.

Exercises

1. Write out in similar fashion a set of rules for subtraction, multiplication, and division.

2. Write an algorithm for expressing an integer in the binary scale in the scale of r.

3. Write algorithms for the elementary arithmetic operations in the binary scale.

4. Write an algorithm for solving the linear algebraic system $a_1x_1 + a_2x_2 = b_1$, $a_3x_1 + a_4x_2 = b_2$.

5. Computer Programming

We shall henceforth assume that the reader has some experience with computer programming, or is currently learning some computer programming language. Much of what follows is meaningful for those who are only acquainted with hand calculation or desk calculators at best, and can be read with profit. Nonetheless, our basic orientation is toward the treatment of problems with the aid of a digital computer. For readers with or without computer experience, we wish to point out that a *machine program* or *computer program* is an algorithm (which is to say, a set of instructions), written in complete detail in a language or code understandable by the machine in question. *Flow charts* are semipictorial representations of the algorithm, usually with instructions or descriptive statements appearing in the English language (or whatever natural language is spoken by the writer) rather than in the machine-understandable or artificial language. Flow charts display in a vivid fashion the branches and repetitive steps in an algorithm of the foregoing nature and are thus an aid to understanding the algorithm. They are often for this reason useful preliminaries to the writing of the machine program. They are similar in structure to the logic trees we used in Chapter One.

For the benefit of readers unfamiliar with flow charts, let us draw one for the problem of alphabetizing the four cards. The result might appear as in Fig. 1. In this diagram, certain conventional symbols are used: An oval denoting an input or output to or from the computer; a diamond enclosing a question, where the answer to the question determines which of several alternative steps to take next; and a rectangle enclosing a step or instruction to be performed unconditionally. The lines with arrows on them indicate the order of "flow" of the steps.

Let us emphasize two important features which we require an algorithm appearing in this volume to possess. First, it must be *deterministic*

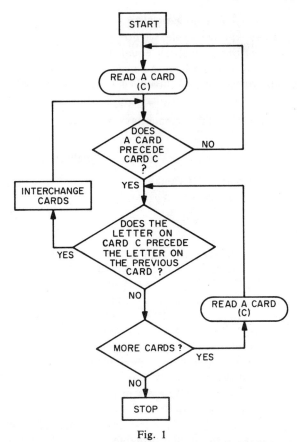

Fig. 1

in the sense that it specifies unambiguously the exact procedure to be followed at each step. It follows that any calculation performed by one calculator or machine can be duplicated by any other calculator or machine (after translating the algorithm to a new language, if necessary), with identical results.* A second important feature of an algorithm is that it must lead to a solution of "any problem of a given kind," rather than to one particular problem only. For example, the foregoing algorithm for alphabetizing will work no matter what four letters appear on the four cards. Similarly, there are algorithms for solving systems of linear algebraic equations, for solving quadratic equations, for performing various operations in calculus, and so on.

In the study of more complex processes we occasionally give up the requirement of a deterministic, explicitly reproducible procedure and allow

* With the proviso that computers with different allowable word length and round-off rules will produce different results in general.

algorithms with chance mechanisms, and we may also waive the requirement of universality. For example, we don't ask for a uniform method for solving all systems of linear algebraic equations with guaranteed numerical accuracy. In this volume we have no need of the sophistication of nondeterministic algorithms. We do, however, emphasize the use of particular algorithms for problems possessing certain structural properties, in addition to presenting general methods.

Exercises

1. Consider the problem of alphabetizing four cards, each containing two letters of the alphabet. The cards are to be put in the usual lexicographic order, so that AZ appears before BA, BA appears before BB, and so on. Assume that the machine which reads the cards can compare only two letters at a time—for example, it can compare the first letters on two cards or the second letters on two cards, but cannot in one step compare a pair of letters on one card with a pair of letters on another card. Write an algorithm and a flow chart for this problem.

2. Modify the algorithm and flow chart of Fig. 1 for the following "bubble sorting" process. A sequence of positive integers is read from cards and stored in the memory of the computer. Each card contains one integer, and the end of the sequence or input file is indicated by a blank card. The first and second numbers are compared, and are exchanged if the first is greater than the second. Now the second (which may originally have been the first) and third are compared and possibly exchanged; then the third and fourth, and so on to the last two numbers. At this point, one *sorting pass* has been completed, and the largest number has "bubbled down" to the last position. A second sorting pass is now made on the sequence consisting of all numbers but the last, after which the next-to-largest number will be in the next-to-last position. A third sorting pass is next made on all numbers but the last two, and so on until $N - 1$ passes have been made, where N is the number of integers in the sequence.

3. Write a FORTRAN or ALGOL program for Exercise 2.

4. If no exchange is made on an entire pass, the sorting is complete and the process can be terminated. Modify the program of Exercise 3 to take advantage of this.

5. The running time of the program can be further decreased by noting that all numbers beyond the last interchange of a pass are already in order. Therefore on the next pass the comparisons would only have

to be made as far as the position of this last interchange. Modify the program to take advantage of this.

6. A method of sorting which sometimes results in increased speed uses an alternating direction of scan. That is, on the first pass the largest number is bubbled down to the last position, N, on the next pass the smallest of the first $(N - 1)$ numbers is bubbled up to the first position, on the third pass the largest of the numbers in positions 2 to $N - 1$ is moved to position $N - 1$, and so on. Find simple numerical examples where this algorithm requires fewer steps, the same number of steps, and more steps than the algorithm of Exercise 4.

7. Write a flow chart and program for the method of Exercise 6.

8. The following method sorts numbers as they are read, rather than first storing all the numbers and then sorting them. Before the first card is read, a *work space* large enough to contain the entire array of numbers is reserved. The integer from the first card is then read and stored in the first position of the work space. As each succeeding integer is read from a card, it is compared with the integers already in the work area and inserted into the work area at the proper place by the following procedure. For example, suppose there are five integers already in the work area and that the new number is larger than the first three but smaller than the fourth. The fourth and fifth numbers in the work area are then shifted to positions five and six, respectively, and the new number is inserted into position four. On the other hand, if the new number is larger than all those already in the work area, it is simply put at the end. Draw a flow chart for this method of sorting.

9. Write a program for the algorithm of Exercise 8.

10. Refer to the "Proceedings of Sort Symposium," *Communications of the Association for Computing Machinery 6* (1963) 194–282, and report on another method of sorting described there.

6. Algorithms for Solving $x = f(x)$

In the light of the discussion in the preceding section, we now see that the method of successive approximations as explained in § 2 and § 3 constitutes an algorithm for the approximate solution of equations of a certain class. In all the examples we gave, the equation under discussion was of the form

$$x = f(x) \tag{1}$$

where the function f varied from one example to another. The method itself consisted in selecting an initial value x_0 which appeared to be reasonably close to the sought-after solution, and then generating successive values sequentially by means of the equations

$$x_1 = f(x_0)$$
$$x_2 = f(x_1)$$
$$x_3 = f(x_2)$$

$$\cdots$$

This sequence of equations can be expressed more concisely in the form

$$x_n = f(x_{n-1}), \qquad n = 1, 2, 3, \cdots \tag{2}$$

Equations such as this are often called *recurrence relations*, since the values of $x_1, x_2, \cdots$ are calculated by a recurrent, or iterative, procedure. To emphasize the similarity of this algorithm to our previous examples, we shall give a flow chart. First, however, let us comment on the question of stopping the process. Recall that in the example of § 2, the equation $x = (x + 1)^{1/2}$, the process continued indefinitely, yielding better and better approximations to the desired solution. In practice, of course, the procedure would be terminated by some criterion for stopping. For example, it could be terminated after a fixed number of steps, or when x_n differs from x_{n-1} by less than 10^{-5}, or 10^{-3}, or whatever is desired. Adopting a "stopping rule" of this latter kind, we can now draw the flow diagram in Fig. 2. Further discussion of stop rules will be given below in § 10.

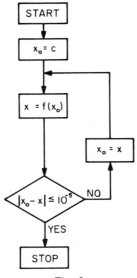

Fig. 2

The example in § 3 brings out the point that the method of successive approximations works in some cases and not in others. This is true of all of the algorithms of analysis as we have noted. This means that a careful preliminary study of the validity of the method used to obtain numerical values is always desirable. These studies constitute the substance of the existence and uniqueness theorems of classical mathematical analysis.

Although this preliminary study is desirable, it is not always possible to carry it through. In some, and indeed far too many, cases, our current level of mathematical analysis is not up to the task of unravelling the complexity of the problem we wish to solve. In these cases, we must use procedures which have worked well in other situations and hope for the best. There used to be a phrase current in football circles: "a pass and a prayer." Nowadays, in computational circles, it might read: "a computer and a prayer."

Frequently and fortunately, however, we possess various tests for the validity of the solution, derived from considerations of the underlying physical process, which we can use as the calculations continue. For example, we may know, in advance, that the solution must be positive, that it must lie in the interval [1, 2], and so on. In some cases, as in the simple examples presented in § 3, the failure of the computational procedure becomes readily apparent, which means that we can stop the calculations without wasting too much time.

The three examples of successive approximations applied to the equation $1 + x = x^2$ show that the success of the method depends critically upon how it is applied. How then do we obtain efficient algorithms in dealing with new types of equations that have not been previously treated? The correct answer is, an might be expected, that this is an art, a magic blend of training, experience, and intuition. There are, nonetheless, certain basic ideas that can be applied successfully in a wide range of cases. One of these is the following: Try to make the approximations reflect some physical operations which are meaningful in the actual process under study.

If the operations are meaningful within the actual physical process, there is a high probability that this will be reflected analytically. Thus, an understanding of the structure of the physical situation underlying the mathematical equations will usually furnish valuable clues.

In our subsequent study of the routing problem, we wish to illustrate how an understanding of the physical process leads to a useful method of successive approximations. This is a theme which runs through the application of mathematics to biology, economics, engineering and physics. Meanwhile, in order to prepare the reader for this, we shall indicate the use of a simple, but extremely powerful, geometric technique for dealing with equations of the form

$$g(x) = 0 \qquad\qquad (3)$$

This method is a very famous one, called the *Newton–Raphson method.* Sometimes, and quite unjustly as far as Raphson is concerned, it is abbreviated and called merely "Newton's method."

We shall assume an elementary knowledge of differential calculus in what follows, namely that required to determine the equation of a tangent to the curve $y = f(x)$ at a specified point.

We confess that it is a bit embarrassing that the simplest examples we can conjure up to illustrate the method of successive approximations require more mathematical training than we require for the subsequent treatment of Eq. (1.1). This illustrates, however, a fundamental point that "elementary" is not synonymous with "simple."

In compensation, however, we feel that it is important for the student both to appreciate the versatility of the method of successive approximations and to become aware of the problems usually encountered in its application.

7. The Newton–Raphson Method : Square Roots

To illustrate the method in one of its simplest and most ancient forms, let us consider the problem of finding the square root of a positive number a. In analytic terms, we wish to find the positive solution of the equation

$$x^2 = a \qquad (1)$$

In Fig. 3, we have a graph of the parabola,

$$y = x^2 - a \qquad (2)$$

Our objective is to determine numerically the coordinates of P, the point $(\sqrt{a}, 0)$. Since (1) does not have the form $x = f(x)$, we cannot immediately apply the algorithm discussed in § 6.

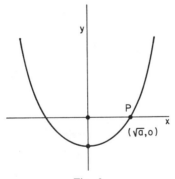

Fig. 3

Suppose we make an initial guess to the x-coordinate of P, say x_1. For example, if $a = 5$, we might guess $x_1 = 3$ (see Fig. 4). How do we find a better estimate?

We suppose that the parabola is approximately, for our purposes, the straight line which is tangent to it at the point Q_1. If the first guess is good enough, this is a reasonable procedure as we shall see. It turns out that the method works well in this case, even if the first guess is not very accurate. This is not to be expected in general. Usually, the success of the method depends critically upon a judicious choice of initial approximation.

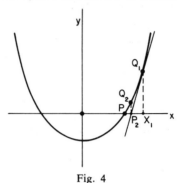

Fig. 4

The equation of the straight line tangent to the parabola at Q_1 is

$$y - (x_1^2 - a) = 2x_1(x - x_1) \tag{3}$$

since the slope of the tangent at Q_1 is $2x_1$. This line hits the x-axis at the point P_2, whose x-coordinate is given by the solution of

$$-(x_1^2 - a) = 2x_1(x - x_1) \tag{4}$$

The x-coordinate of the point P_2 is thus

$$x_2 = \frac{a + x_1^2}{2x_1}. \tag{5}$$

We now repeat this procedure. At the point Q_2, with the coordinates $(x_2, x_2^2 - a)$, we again draw the tangent. Its equation is

$$y - (x_2^2 - a) = 2x_2(x - x_2) \tag{6}$$

and its intersection with the x-axis is at the point

$$x_3 = \frac{a + x_2^2}{2x_2} \tag{7}$$

We now continue in this fashion, generating a sequence of values $x_1, x_2,$ $x_3, \cdots$.

It is easy to see geometrically that x_2 exceeds $\sqrt{a}$ and that generally

$$\sqrt{a} < \cdots < x_3 < x_2 \tag{8}$$

We urge the reader to establish this sequence of estimates analytically by means of the recurrence relation,

$$x_n = \frac{a + x_{n-1}^2}{2x_{n-1}} \tag{9}$$

Once again, the fundamental result concerning the convergence of infinite sequences assures us that x_n converges as n increases without limit and that $\lim_{n \to \infty} x_n = \sqrt{a}$. Although we shall provide a number of exercises at the end of this section illustrating various interesting aspects of the convergence of the sequence $\{x_n\}$, we shall not discuss this property in the text for reasons previously discussed. In the first place, the concept of convergence requires a higher level of mathematical education and, more importantly, of mathematical sophistication than that demanded elsewhere in the book. Secondly, and quite fortunately, let us emphasize that the method employed to treat the equation in (1.1) yields the solution after a *finite* number of steps. Consequently, no discussion of infinite sequences is necessary for a completely rigorous treatment of the problems of principal interest to us.

Exercises

1. Do we have convergence if $x_1 < \sqrt{a}$? Is the sequence monotone decreasing as before? Discuss the two cases $0 < x_1 < \sqrt{a}$, $-\infty < x_1 < -\sqrt{a}$. Does the sequence ever converge to the other root, $-\sqrt{a}$?

2. Show that if we wish to use the same procedure to find the cube root of a, i. e., to find the positive solution of $x^3 = a$, the relations corresponding to (5) and (7) are

    $$x_2 = \frac{2x_1}{3} + \frac{a}{3x_1^2} \qquad x_3 = \frac{2x_2}{3} + \frac{a}{3x_2^2}$$

    and so on. If $x_1 > 0$, is the sequence monotone increasing or decreasing? Is there always convergence, regardless of the choice of x_1 ($x_1 = 0$ excepted)?

3. Show that if the original equation is $f(x) = 0$, the relations corresponding to (5) and (7) are

    $$x_2 = x_1 - \frac{f(x_1)}{f'(x_1)} \qquad x_3 = x_2 - \frac{f(x_2)}{f'(x_2)}$$

    and so on.

4. How would one extend the Newton–Raphson technique to treat the simultaneous system $f(x_1, x_2) = 0$, $g(x_1, x_2) = 0$? (A treatment of this question can be found in most standard textbooks on numerical analysis.)

5. If the line $y = x$ and the curve $y = f(x)$ are graphed, the graphs intersect at the point (r, r) if r is a root of $x = f(x)$. Draw a hypothetical graph. Show that for any x_{n-1}, $x_{n-1} \neq r$, if a horizontal line is drawn through the point $(x_{n-1}, f(x_{n-1}))$, this line will intersect the line $y = x$ at the point (x_n, x_n), where x_n is the next approximation given by Eq. (6.2). In this way, the method of successive approximations has a simple geometrical representation.

6. (For students who have studied elementary calculus.) Show by a graph that if the slope of the curve $y = f(x)$ at $x = r$ satisfies the condition $|f'(r)| < 1$, then the successive iterates x_n will converge to r provided the initial approximation is sufficiently close to r to begin with. Show graphically that if $|f'(r)| > 1$, the iterates x_n cannot converge.

7. Aitken's delta-squared process is a method for speeding up convergence of an iterative process. If x_1, x_2, x_3, etc., is a convergent sequence, Aitken's method consists in computing x_1^*, x_2^*, etc., from the formula

$$x_n^* = \frac{x_n x_{n+2} - (x_{n+1})^2}{x_{n+2} - 2x_{n+1} + x_n}$$

Suppose $x_1 = 1$, $x_2 = \frac{1}{2}$, $\cdots$, $x_n = 1/n$, $\cdots$. Compute x_1^* and x_2^*. Prove that $x_n^* = x_{2(n+1)}$.

8. An Example

To illustrate the technique, consider the problem of finding $\sqrt{5}$. Take $x_1 = 3$. Then

$$x_2 = \frac{5 + x_1^2}{2x_1} = \frac{5 + 9}{6} = \frac{14}{6} = \frac{7}{3} = 2.3333$$

$$x_3 = \frac{5 + x_2^2}{2x_2} = \frac{5 + 49/9}{14/3} = \frac{45 + 49}{42} = \frac{94}{42} = 2.238 \qquad (1)$$

$$x_4 = 2.236068811$$

We see that x_4 yields $\sqrt{5}$ to five significant figures. To ten significant figures

$$\sqrt{5} = 2.236067978 \qquad (2)$$

The student with a desk calculator can verify that

$$x_5 = 2.236067979$$
$$x_6 = 2.236067978$$

(3)

Exercises

1. Examine the approximations x_4, x_5, x_6. How many significant figures do we obtain at each step?

2. Carry out the same calculation for $\sqrt{2}$, $\sqrt{3}$, and again observe the number of significant figures obtained at each step, using values of $\sqrt{2}$, $\sqrt{3}$ accurate to twenty significant figures for comparison purposes.

3. Show that the error, the value $|x_n - \sqrt{a}|$, is essentially squared at each step of the calculation by proving that

$$|x_{n+1} - \sqrt{a}| \leq \frac{(x_n - \sqrt{a})^2}{\sqrt{a}}$$

Hint: $x_{n+1} - \sqrt{a} = a/2x_n + x_n/2 - \sqrt{a} = (x_n - \sqrt{a})^2/2x_n$.

4. Establish a similar inequality for the general case,

$$x_{n+1} = x_n - \frac{(f(x_n) - a)}{f'(x_n)}$$

where x_n is an approximation to the solution of $f(x) = a$.

9. Enter the Digital Computer

The desk calculator and even plain old pencil and paper are extremely useful in dealing with problems of the foregoing type. They do require, however, patience and scrupulous attention to detail over long periods of time, neither of which are outstanding characteristics of the human being. If we enlarge the problem by considering many equations of this type, or if we consider simultaneous equations, we find that the requirements of time and accuracy overwhelm us.

Can we give these tasks over to the digital computer and save our emotional and intellectual energy for other endeavors which require the human touch?

It turns out that we can. The reason for this is that the determination of x_2 given x_1, x_3 given x_2, and so on, requires only the basic arithmetic operations and a set of simple instructions embodied in the formula

$$x_{n+1} = \frac{a + x_n^2}{2x_n}, \qquad n = 0, 1, 2, \cdots. \tag{1}$$

This is an ideal set up for a digital computer.

In the following sections we will discuss some aspects of the numerical determination of the sequence $\{x_n\}$ using a digital computer.

10. Stop Rules

Suppose that we wish to calculate $\sqrt{a}$ using the sequence

$$x_{n+1} = \frac{a + x_n^2}{2x_n} \tag{1}$$

with $x_0 = c$. How many terms in the sequence should we calculate, and how do we know when to stop? Another way of phrasing this question is the following: What instructions do we give to the computer to end the calculation?

There are several alternatives:

1. Fix the total number of approximations computed in an a priori fashion: $n = 1, 2, 3, 4, 5$, or $n = 1, 2, \cdots, 10$.
2. Use an adaptive procedure: At each stage, calculate the quantity $x_n^2 - a$. Stop when $|x_n^2 - a| < \epsilon$, where ϵ is a preassigned quantity.
3. Use an alternative adaptive procedure: At each stage, calculate the percentage change $|100(x_n - x_{n-1})/x_{n-1}|$. Stop when this quantity is less than or equal to ϵ, where ϵ is a preassigned quantity.

In the case of the routing Eq. (1.1), we can use a simple stop rule: Cease the calculation when the solution is obtained. We can employ this command in this case because we know in advance that not more than a certain number of stages are needed. We will discuss this in greater detail below. Once again, let us emphasize that convergence of the sequence of approximations in a finite number of stages is quite exceptional. In general, stop rules of the foregoing type are required.

11. Flow Charts for the Determination of $\sqrt{a}$

Let us now present two flow charts for the determination of $\sqrt{a}$ based upon the foregoing recurrence relation. Each illustrates a different stop rule.

Since the details of a computer program vary from machine to machine and language to language, we will not exhibit any specific program.

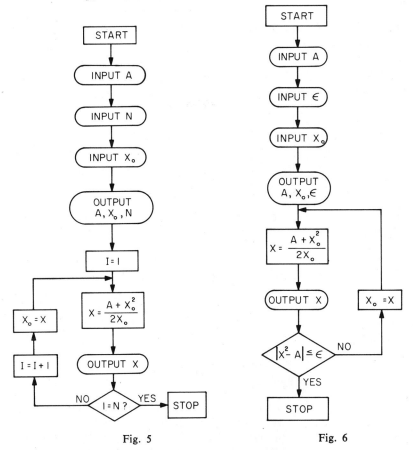

Fig. 5   Fig. 6

12. What Initial Approximation?

In the calculations we have so far discussed, the choice of an initial approximation has been of importance insofar as convergence of the sequence of approximations is concerned. Apart from that, we have not worried about the choice. It is clear, from experiments with the computer, or from "common sense," that the choice of the value of c, the initial approximation, determines the rapidity with which the approximations converge to $\sqrt{a}$. This is of rather slight importance in this particular calculation where the total time is measured in milliseconds. In longer calculations where the time consumed is of the order of hours or tens of hours, this question of a suitable starting point is vital. In our treatment of the routing equation, we shall indicate a number of ways of obtaining good initial approximations, as well as techniques for accelerating convergence.

In the meantime, we provide below a set of Exercises which illustrate the foregoing remarks.

Exercises

1. Consider the sequence $x_{n+1} = x_n/2 + a/2x_n$, $x_0 = c$. Take $a = 2$, and evaluate $|x_5 - \sqrt{2}|$ for $c = 0.1, 0.5, 1, 2, 5, 10$.

2. Take $a = 2, 3, 5, 7, 10$, and calculate the quantity $|x_5 - \sqrt{a}|$ for each value of a. For each of the foregoing values of c, calculate $\max_a |x_5 - \sqrt{a}|$. Try a range of values of c and estimate the value of c which yields $\min_c \max_a |x_5 - \sqrt{a}|$.

3. Carry out the same type of experimentation for the sequence $\{x_n\}$ defined by $x_{n+1} = 2x_n/3 + a/3x_n^2$, $x_0 = c$.

13. An Approximation Method of Steinhaus

As another example of the use of simple geometric ideas to generate a sequence of approximations, let us consider the problem of solving two simultaneous linear algebraic equations

$$L_1: \quad a_1 x + a_2 y + a_3 = 0$$
$$L_2: \quad b_1 x + b_2 y + b_3 = 0 \tag{1}$$

We can, of course, use Cramer's rule (solution by determinants), or we can use an elimination technique. Both of these methods carry over to higher-dimensional cases in theory. In practice, the elimination technique used with a *soupçon* of ingenuity is generally successful, while Cramer's rule is completely unusable for numerical purposes.

Let us here, however, present an elegant approximation method due to Steinhaus which will furnish a number of interesting and illuminating exercises (see Fig. 7). Let us start with a point P_1 on L_2, say (x_1, y_1), and drop a perpendicular to L_1. Call the foot of the perpendicular $P_2 = (x_2, y_2)$, and from this point drop a perpendicular to L_2, and so on. It is clear that $\overline{P_1P_2} > \overline{P_2P_3} > \overline{P_3P_4}$ and that the sequences $(P_1, P_3, \cdots)$, $(P_2, P_4, \cdots)$, con-

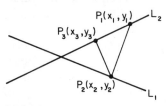

Fig. 7

verge to P. Observe that this method is iterative whereas Cramer's rule is not.

Exercises

1. Write the equations expressing x_2, y_2 in terms of x_1, y_1, the equations for x_3, y_3 in terms of x_2, y_2, and generally x_{2n}, y_{2n} in terms of x_{2n-1}, y_{2n-1}, and x_{2n+1}, y_{2n+1} in terms of x_{2n}, y_{2n}.

2. Write a flow chart for the calculation of x_{2n}, y_{2n} and x_{2n+1}, y_{2n+1}, $n = 1, 2, \cdots$

3. Consider the following stop rules:

   (a) Stop when $|(x_{2n} - x_{2n-1})/x_{2n}| + |(y_{2n} - y_{2n-1})/y_{2n}| \leq \epsilon$
   (b) Stop when $(a_1x_n + a_2y_n + a_3)^2 + (b_1x_n + b_2y_n + b_3)^2 \leq \epsilon$
   (c) Stop when $n = N$, where N is a preassigned number.

   What are the advantages and disadvantages of each rule?

4. Suppose that we have three simultaneous equations:

$$P_1: \quad a_1x + a_2y + a_3z + a_4 = 0$$
$$P_2: \quad b_1x + b_2y + b_3z + b_4 = 0$$
$$P_3: \quad c_1x + c_2y + c_3z + c_4 = 0$$

   Let us take a point $Q_1(x_1, y_1, z_1)$ on the plane P_1 and drop a perpendicular to P_2. Call the foot of the perpendicular $Q_2(x_2, y_2, z_2)$. From Q_2 drop a perpendicular to P_3 and so on. Write down the equations connecting (x_n, y_n, z_n) and $(x_{n-1}, y_{n-1}, z_{n-1})$, and construct a flow chart, using various stop rules.

5. Is it best to keep a fixed order, $P_1 \to P_2 \to P_3 \to P_1$? What would encourage one to vary the order, and what then would the flow chart look like?

14. Successive Approximations for a Simple Routing Problem

We shall now show how the method of successive approximations can be applied to the equations we have derived for the maps in Chapter One, and, in general, to equations of the form

$$f_i = \min_{j \neq i} (t_{ij} + f_j), \qquad i = 1, 2, \cdots, N-1,$$
$$f_N = 0 \tag{1}$$

To illustrate, let us begin with a simple special case of (1), for example, a set of equations obtained from the map of Fig. 8. Suppose $t_{21} =$

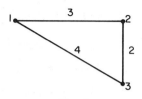

Fig. 8

$t_{12} = 3$, $t_{32} = t_{23} = 2$, $t_{31} = t_{13} = 4$. The corresponding equations are

$$f_1 = \min_{j=2,3} (t_{1j} + f_j) = \min (3 + f_2, \, 4 + f_3)$$
$$f_2 = \min_{j=1,3} (t_{2j} + f_j) = \min (3 + f_1, \, 2 + f_3) \qquad (2)$$
$$f_3 = 0$$

We begin by making a first *guess* for the values of f_1, f_2, and f_3. For example, suppose one makes the rather wild guess

$$f_1^{(1)} = 8, \quad f_2^{(1)} = 0, \quad f_3^{(1)} = 0 \qquad (3)$$

The number one in each parenthesis is a *superscript*,* and indicates that this is the starting approximation, or "first approximation." These values are sometimes called the "zero-th approximation" and denoted by $f_1^{(0)}$, $f_2^{(0)}$, $f_3^{(0)}$, in which case the quantities $f_1^{(1)}$, $f_2^{(1)}$, $f_3^{(1)}$ are called the first approximation, meaning the values obtained after one step, or one iteration. It does not matter greatly which notation is selected, and as we have remarked, we shall use both in this book. Some digital computer languages do not allow the use of zero in such a way.

The first approximate values are now substituted into the right members in (2), and the minima calculated, yielding values of a second approximation, as follows

$$f_1^{(2)} = \min (3 + f_2^{(1)}, \, 4 + f_3^{(1)}) = \min (3 + 0, \, 4 + 0) = 3,$$
$$f_2^{(2)} = \min (3 + f_1^{(1)}, \, 2 + f_3^{(1)}) = \min (3 + 8, \, 2 + 0) = 2, \qquad (4)$$
$$f_3^{(2)} = 0.$$

These newly calculated values are again substituted into the right members of (2), yielding

$$f_1^{(3)} = \min (3 + 2, \, 4 + 0) = 4,$$
$$f_2^{(3)} = \min (3 + 3, \, 2 + 0) = 2, \qquad (5)$$
$$f_3^{(3)} = 0.$$

This process could now be repeated over and over again. However, one finds that the third and fourth approximations are the same. Obviously

* Superscripts have been used reluctantly to avoid double subscripts.

there is no point in continuing these operations, since no further change can occur. We have obtained a set of numbers which satisfy (2).

It is easy to see that the values obtained are in fact the desired minimum times required to reach point 3, namely $f_1 = 4$, $f_2 = 2$, $f_3 = 0$. Thus the approximations, in the present instance, converge to the exact solution in a finite number of steps, in fact, in a very few steps. This is in sharp contrast to what happened in the problem of § 7. There the sequence of approximations converged to the solution $\sqrt{5}$, in the sense that the difference between x_n and $\sqrt{5}$ became ever closer to zero, but for no value of n was x_n exactly equal to $\sqrt{5}$, nor could it be, since each x_n must be a rational number if x_1 is rational, while $\sqrt{5}$ is irrational.

15. The Map of Fig. 3 of Chapter One

The same kind of procedure can be applied to the equations which describe Fig. 3 of Chapter One. The size of the set of equations suggests strongly that we give some serious thought as to how to select the initial guess, since a poor choice may result in very slow convergence. It is reasonable to suppose that results will be significantly better if our first approximation is not too far off. For example, we might take $f_i^{(1)}$ to be the time to reach 8 from i following a route which goes directly up, if possible, and otherwise directly to the right. A glance at Fig. 3 shows that this is equivalent to choosing the route of shortest distance (rather than time). The times obtained should be fairly close to the best, since any route of least time would be expected to move generally upward and to the right. We shall call such a rule, which tells us how to select the initial approximation, an *initial policy*. This initial policy is usually based upon experience or knowledge of the exact solution of a simpler problem. In the present case, this policy dictates the following choice

$$
\begin{aligned}
f_8^{(1)} &= 0 \\
f_7^{(1)} &= \text{time to reach 8 on route } 7, 8 & &= 43 \\
f_6^{(1)} &= \quad '' \quad '' \quad '' \quad '' \quad '' \quad '' \quad 6, 7, 8 & &= 90 \\
f_5^{(1)} &= \quad '' \quad '' \quad '' \quad '' \quad '' \quad '' \quad 5, 7, 8 & &= 105 \\
f_4^{(1)} &= \quad '' \quad '' \quad '' \quad '' \quad '' \quad '' \quad 4, 5, 7, 8 & &= 121 \\
f_3^{(1)} &= \quad '' \quad '' \quad '' \quad '' \quad '' \quad '' \quad 3, 5, 7, 8 & &= 152 \\
f_2^{(1)} &= \quad '' \quad '' \quad '' \quad '' \quad '' \quad '' \quad 2, 3, 5, 7, 8 & &= 179 \\
f_1^{(1)} &= \quad '' \quad '' \quad '' \quad '' \quad '' \quad '' \quad 1, 3, 5, 7, 8 & &= 198 \\
f_0^{(1)} &= \quad '' \quad '' \quad '' \quad '' \quad '' \quad '' \quad 0, 1, 3, 5, 7, 8 & &= 226
\end{aligned}
\tag{1}
$$

Some calculation is required to obtain these values of $f_i^{(1)}$. However,

the fact that we have made a choice of an initial policy enables us to avoid
a complete enumeration of paths or the drawing of a tree diagram. Instead
we form exactly one path for each starting point. The policy tells us where
to go next at each intersection. For large maps, this will result in a great
saving of computer time.

We shall now show how to compute the higher approximations in a
convenient and systematic way. Later in the Chapter we shall show how
to mechanize the calculation for a computer, but at the moment we shall
carry out the calculations by hand, or perhaps with the aid of a desk calculator. First of all, for convenience we reproduce Table 10 of Chapter
One (this time arranged so that the "from" vertex is at the top and the "to"
vertex on the left side). The result is given in Table 1. Now we calculate

TABLE 1

MATRIX OF TIMES FOR FIG. 3 CHAPTER ONE

From To	0	1	2	3	4	5	6	7	8
0	0	28	40						
1	28	0		46					
2	50		0	27	37				
3		46	27	0		17			
4			52		0	26	118		
5				47	16	0		82	
6					103		0	47	
7						62	47	0	33
8								43	0

$f_0^{(2)}$ from the formula, (cf., the example in § 14)

$$f_0^{(2)} = \min (t_{01} + f_1^{(1)}, \ t_{02} + f_2^{(1)}) \tag{2}$$

The values of t_{01} and t_{02} from Table 1 are 28 and 50, and $f_1^{(1)} = 198$, $f_2^{(1)} = 179$. Consequently, $f_0^{(2)} = 226$. Next,

$$
\begin{aligned}
f_1^{(2)} &= \min_{j=0,3} (t_{ij} + f_j^{(1)}) \\
&= \min (t_{10} + f_0^{(1)}, \ t_{13} + f_3^{(1)}) \\
&= \min (28 + 226, \ 46 + 152) \\
&= 198
\end{aligned}
\tag{3}
$$

We continue in this way until all of the $f_i^{(2)}$, $i = 0, 1, 2, \cdots, 7$, have been
found.

The procedure can be speeded up by making a strip of paper on which the values of the $f_i^{(1)}$ are listed in a column. Suppose this column* is placed on top of Table 1, next to the column headed 0, as shown in Fig. 9. Then in the two columns we have t_{0j} and $f_j^{(1)}$. By adding adjacent items, we get $t_{0j} + f_j^{(1)}$, and the smallest of these is $f_0^{(2)}$, as we see from (2). We must consider only those values of j, different from 0, for which there is a street from 0 to j, but happily this is apparent from Fig. 9 itself, since the lack of a street connection is indicated by a blank space in the t_{0j} column.

To	From 0	$f_j^{(1)}$
0	0	226
1	28	198
2	50	179
3		152
4		121
5		105
6		90
7		43
8		0

Fig. 9

In the same way, placing the $f_j^{(1)}$ strip next to the t_{1j} column, we can easily obtain $f_1^{(2)}$, placing it next to the t_{2j} column we obtain $f_2^{(2)}$ (see Fig. 10), and so on. The results can be recorded as in Table 2, in which we include an extra column to tell which value of j produced the minimum. Thus from Fig. 9, we see that the minimum of 226 is attained for $j = 1$, and from Fig. 10 the minimum of 173 is attained for $j = 4$.

The numbers $f_j^{(2)}$ can now be written in a column on a strip of paper, and placed over Table 1 just as before in order to calculate a third set of

To	From 0	1	2	$f_j^{(1)}$
0			40	226
1				198
2			0	179
3			27	152
4			52	121
5				105
6				90
7				43
8				0

Fig. 10

* If the data in Table 1 are arranged with the "from" vertex on the left and the "to" vertex at the top, the entries $f_i^{(1)}$ should be listed in a row instead of a column.

TABLE 2

j	$f_j^{(1)}$	Minimizing Vertex	$f_j^{(2)}$
0	226	1	226
1	198	3	198
2	179	4	173
3	152	5	152
4	121	5	121
5	105	7	105
6	90	7	90
7	43	8	43
8	0	—	0

numbers $f_j^{(3)}$. These in turn can be used to find a fourth approximation, and so on. The results of this procedure are given in Table 3. It is seen that once again we obtain convergence in a finite number of steps. Since the values $f_i^{(3)}$ and $f_i^{(4)}$ are equal for each i, they satisfy the equations for the minimal times and thus give the minimal times to reach 8, assuming for the moment the uniqueness theorem established in Chapter Four.

TABLE 3

INITIAL POLICY: UP IF POSSIBLE, OTHERWISE RIGHT

j	$f_j^{(1)}$	Min j	$f_j^{(2)}$	Min j	$f_j^{(3)}$	Min j	$f_j^{(4)}$
0	226	1	226	2	223	2	223
1	198	3	198	3	198	3	198
2	179	4	173	4	173	4	173
3	152	5	152	5	152	5	152
4	121	5	121	5	121	5	121
5	105	7	105	7	105	7	105
6	90	7	90	7	90	7	90
7	43	8	43	8	43	8	43
8	0	—	0	—	0	—	0

16. Determination of Optimal Policy

We have obtained *a* solution to the system of equations

$$f_i = \min_{j \neq i} [t_{ij} + f_j], \qquad i = 0, 1, \cdots, 7$$
$$f_8 = 0 \tag{1}$$

Let us assume for the moment that it is the only solution and see how it can be used to solve the original routing problem. From the relation

$$f_0 = \min_{j \neq 0} [t_{0j} + f_j] \qquad (2)$$

and the values of f_j we have obtained, we see that the unique minimum value is determined for $j = 2$. Hence we go from 0 to 2. The relation

$$f_2 = \min_{j \neq 2} [t_{2j} + f_j] \qquad (3)$$

similarly yields the minimizing value $j = 4$. Hence, we proceed from 2 to 4. In the same fashion, we see that from 4 we go to 5, from 5 to 7, and finally from 7 to 8. The optimal path is thus $[0, 2, 4, 5, 7, 8]$.

What is interesting then about (1) is that the single equation provides two functions: the minimal time function f_i, and the policy function j_i which tells to what point j to go next from any point i. The minimum time function is clearly uniquely determined. However, the policy function may well be multivalued. This is to say, there may be several paths from the same point which yield the desired minimum time. This occurs when we have a tie in (1), namely $t_{ij} + f_j = t_{ik} + f_k$, both yielding the minimum value. Then $j_i = j$ and k.

A policy which produces the minimum time function is called an *optimal policy*.

We see how vital uniqueness of solution is, since the optimal policy j_i is determined by minimizing the function $t_{ij} + f_j$ over j. If there were several solutions for the f_i and we used the wrong one we could easily obtain spurious values for j_i. In Chapter Four we will establish the desired uniqueness of solution of Eq. (1.1).

Exercise

1. Using Table 3, determine the optimal paths from 2, 3, 4, 5, 6, 7.

17. A Different Initial Policy

There are, naturally, many other possible choices for the initial policy or guess. In order to illustrate what happens if the first guess is not very effective, let us work through the calculation when the initial policy for choosing a path is the following: go down if possible otherwise to the right if possible, and when the right edge of the map is reached, go up to 8. The first guess is then as follows:

Start at	Path is	Time is
0	0, 2, 4, 6, 7, 8	295
1	1, 0, 2, 4, 6, 7, 8	323
2	2, 4, 6, 7, 8	245
3	3, 2, 4, 6, 7, 8	272
4	4, 6, 7, 8	193
5	5, 4, 6, 7, 8	219
6	6, 7, 8	90
7	7, 8	43

The first two steps in the calculation are given in Table 4. We leave it to the reader to check these steps and to complete the calculation, showing that $f_i^{(5)}$ and $f_i^{(6)}$ are the same, so that the process converges, but in two more steps than before.

What this illustrates once again is that the efficacy of the method of successive approximations depends critically upon both how it is applied and the choice of an initial approximation.

TABLE 4

Initial Policy: Down If Possible, Right, Up

j	$f_j^{(1)}$	Min j	$f_j^{(2)}$	Min j	$f_j^{(3)}$	Min j	$f_j^{(4)}$	Min j	$f_j^{(5)}$	Min j	$f_j^{(6)}$
0	295	2	295	2	295						
1	323	3	318	3	312						
2	245	4	245	4	245						
3	272	5	266	5	152						
4	193	6	193	5	121						
5	219	7	105	7	105						
6	90	7	90	7	90						
7	43	8	43	8	43						
8	0	—	0	—	0						

There are other good choices of initial policy suggested by the nature of this problem, some of which are listed in the problems below.

Exercises

Work through the method of successive approximations for each of the following choices of initial policy:

1. $f_j^{(1)}$ is the time required to go from j to 8 on the path having the fewest turns. No turn is to be counted at the starting vertex j.

2. $f_j^{(1)}$ is the time required to go from j to 8 on the path having the smallest total number of stop signs and stop lights. If several routes have the same total, choose any one, say the most northerly one.

18. Successive Approximations for A General Map

The method of successive approximations used in the foregoing pages can be applied to solve the quickest route problem for any map. If the map has N vertices, we have seen that the minimal times from i to N, f_i, satisfy the equations

$$f_i = \min_{j \neq i} (t_{ij} + f_j) \qquad i = 1, 2, \cdots, N-1 ,$$
$$f_N = 0 \tag{1}$$

where one minimizes over all j which can be reached directly from i. The general method of successive approximations now proceeds according to these rules:

1. Choose an initial guess $f_i^{(1)}$ ($i = 1, 2, \cdots, N-1$) in any way, and choose $f_N^{(1)} = 0$.

2. Compute $f_i^{(2)}$ from $f_i^{(1)}$, and so on, generally $f_i^{(k)}$ from $f_i^{(k-1)}$, using the recurrence relations

$$f_i^{(k)} = \min_{j \neq i} (t_{ij} + f_j^{(k-1)})$$
$$f_N^{(k)} = 0 \qquad k = 2, 3, \cdots . \tag{2}$$

Let us apply the method to the map of Fig. 11, of Chapter One, for which we found that the method of enumeration becomes quite cumbersome. In § 15 of Chapter One we gave some arithmetical characterizations of what we mean by "cumbersome."

The initial policy is to choose a route that goes up if possible, otherwise to the right if possible, and down if neither up nor right is possible. Using the times from Table 11 of Chapter One, reproduced in Table 5 below, the values $f_i^{(1)}$ are easily calculated.

$$f_{11}^{(1)} = 0$$

$f_{10}^{(1)}$ = time to reach 11 on route 10, 11   = 79

$f_9^{(1)}$ = ″ ″ ″ ″ ″ ″ 9, 10, 11   = 113

$f_8^{(1)}$ = ″ ″ ″ ″ ″ ″ 8, 9, 10, 11   = 169

$f_7^{(1)}$ = ″ ″ ″ ″ ″ ″ 7, 11   = 43

$f_6^{(1)}$ = ″ ″ ″ ″ ″ ″ 6, 7, 11   = 90

$f_5^{(1)}$ = ″ ″ ″ ″ ″ ″ 5, 10, 11   = 111

$f_4^{(1)}$ = ″ ″ ″ ″ ″ ″ 4, 5, 10, 11   = 127

(3)

$$f_3^{(1)} = \quad '' \quad '' \quad '' \quad '' \quad '' \quad '' \quad 3, 9, 10, 11 \quad = 151$$

$$f_2^{(1)} = \quad '' \quad '' \quad '' \quad '' \quad '' \quad '' \quad 2, 3, 9, 10, 11 \quad = 178$$

$$f_1^{(1)} = \quad '' \quad '' \quad '' \quad '' \quad '' \quad '' \quad 1, 8, 9, 10, 11 \quad = 192$$

$$f_0^{(1)} = \quad '' \quad '' \quad '' \quad '' \quad '' \quad '' \quad 0, 1, 8, 9, 10, 11 = 220$$

TABLE 5

TIMES FOR FIG. 11 OF CHAPTER ONE

From\To	0	1	2	3	4	5	6	7	8	9	10	11
0	0	28	40									
1	28	0		46					33			
2	50		0	27	37							
3		46	27	0		17				48		
4			52		0	26	118					
5				47	16	0		82			32	
6					103		0	47				
7						62	47	0				33
8		23							0	56		
9				38					56	0	34	
10						32				34	0	89
11								43			79	0

TABLE 6

SUCCESSIVE APPROXIMATIONS FOR FIG. 11

j	$f_j^{(1)}$	Min Vertex	$f_j^{(2)}$	Min Vertex	$f_j^{(3)}$	Min Vertex	$f_j^{(4)}$	Min Vertex	$f_j^{(5)}$
0	220	1	220	1	220	1	220	1	220
1	192	8	192	8	192	8	192	8	192
2	178	3	178	3	178	4	173	4	173
3	151	9	151	9	151	9	151	9	151
4	127	5	127	5	121	5	121	5	121
5	111	7	105	7	105	7	105	7	105
6	90	7	90	7	90	7	90	7	90
7	43	11	43	11	43	11	43	11	43
8	169	9	169	9	169	9	169	9	169
9	113	10	113	10	113	10	113	10	113
10	79	11	79	11	79	11	79	11	79
11	0	—	0	—	0	—	0	—	0

The computation using recurrence relations now proceeds as before, with the results indicated in Table 6 above. It is a striking fact that the process converges in only one step more than for the smaller map, and that the total calculation time is not even doubled. Table 6 can be constructed by hand from Table 5, in four or five minutes. Contrast this with the lengthy and tedious enumerative process of Chapter One! Nevertheless, this preliminary effort was not wasted. Only by comparing a pedestrian approach with a sophisticated method do we appreciate the importance of a new mathematical technique.

For maps of one hundred vertices or less, we remain in the realm of hand calculation. The solution by a novice using the recurrence relations given above may take one hour or so. For maps of several thousand vertices, it is a matter of a simple computer calculation, requiring a few minutes at the present time. With a knowledge of the times required to perform an addition, it is not difficult to estimate the total time needed to obtain the solution. Within a few years, it will be possible to treat maps containing several hundred thousand or several million vertices in the same straightforward fashion.

Nonetheless, there are interesting new mathematical questions which arise when large maps are considered. We will discuss some of these in Chapter Three and also indicate there why the routing problem is far more than either an entertaining puzzle or merely an instructive mathematical problem for education in the use of the computer. Let us emphasize that the domain of large-scale problems is openended. As computers become more powerful and mathematicians become more familiar with the nature of the problems, we will be able to penetrate further and further and label many more problems "tame." But there will always be *terra incognita* within the mathematical map of this area, and the natives will not always be friendly.

It is, of course, true that part of the efficiency of this method is due to the fact that our initial policy was very close to the optimal choice. If one's initial guess is not very good, the process may be much longer. This is illustrated in the exercises below.

Exercises

Solve the quickest route problem for each of the following initial policies:

1. The time to go from j to 11 by the path with the fewest stop signs or stop lights is chosen as $f_j^{(1)}$.

2. The time to go from j to 11 by the shortest (in feet) path is chosen as $f_j^{(1)}$. See Table 7 of Chapter One for the distance.

3. Suppose that we wish to find the quickest route from one fixed vertex, 1, to every other vertex. Then as shown in the Exercise in § 13 of Chapter One the basic equations are $f_i = \min_{j \neq i} (t_{ji} + f_j)$, $f_1 = 0$. Formulate equations for successive approximations and use these and a suitably chosen initial guess to find the quickest route from vertex 1 to every other vertex for the map with data as in Table 1.

19. Computer Program

Let us briefly outline a computer method for performing the calculations indicated by Eq. (18.2).

Step 1. Read into computer storage the number N of vertices, the matrix of times, and the initial approximation vector. We shall here denote these times by $T(I, J)$ and the elements of the initial vector by $F(J)$. For the sake of simplicity it can at first be assumed that there is a link connecting every pair of vertices so that $T(I, J)$ exists for $I, J = 1, 2, \cdots, N$. Ways of removing this restriction will be suggested below.

Step 2. For $I = 1, 2, \cdots, N - 1$, calculate

$$\text{FMIN}(I) = \min_{J \neq I} [T(I, J) + F(J)], \tag{1}$$

and NEXT(I), the vertex J which yields the minimum in Eq. (1).

Step 3. Set $F(I) = \text{FMIN}(I)$ for $I = 1, 2, \cdots, N - 1$. If there is no change here, stop, otherwise return to step 2.

We leave it to the reader to construct a flow chart and detailed program.

20. The Connection Matrix and A Second Programming Method

If the quantity t_{ij} does not exist for every distinct pair of vertices, the program in the preceding section cannot be used. There are several ways of bypassing this difficulty, and indeed even of turning it to advantage. In this section we shall give one such method which enables us to introduce the important concept of the *matrix associated with* a graph. Later, we shall suggest several other programming methods which will be better adapted to many practical problems.

Let a graph be given, with vertices $1, 2, \cdots, N$. Let the quantity c_{ij} be set equal to *one* if there is a directed link going from vertex i to vertex j, and let c_{ij} be set equal to *zero* if there is not. The matrix C with entry

c_{ij} in the ith row and jth column will be called the *matrix associated with the graph*, or *connection* or *adjacency* or *connectivity matrix* of the graph.

For example, for the graph in Fig. 11, the connection matrix is given

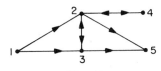

Fig. 11

TABLE 7

$$C = \begin{bmatrix} 0 & 1 & 1 & 0 & 0 \\ 0 & 0 & 1 & 1 & 1 \\ 0 & 1 & 0 & 0 & 1 \\ 0 & 1 & 0 & 0 & 0 \\ 0 & 0 & 0 & 0 & 0 \end{bmatrix}$$

in Table 7. In the figure, arrows indicate the direction in which a link may be traversed, and some of the links are assumed to be oneway. In this example no link leaves vertex 5, and because of this the last row in C consists entirely of zeros.*

There is a close relation between the matrix C and the matrix T of times t_{ij}. In fact, if an entry t_{ij} appears in T, the entry c_{ij} (with the same i and j) must be 1, since there is a link from i to j. For i and j for which t_{ij} does not exist, c_{ij} must be 0. In other words, C can be obtained from T by changing each nonzero entry to one and filling each blank with the entry zero. For example, from Table 5 we readily see that the connection matrix for Fig. 11 of Chapter One is as follows:

$$
\begin{array}{cccccccccccc}
0 & 1 & 1 & 0 & 0 & 0 & 0 & 0 & 0 & 0 & 0 & 0 \\
1 & 0 & 0 & 1 & 0 & 0 & 0 & 0 & 1 & 0 & 0 & 0 \\
1 & 0 & 0 & 1 & 1 & 0 & 0 & 0 & 0 & 0 & 0 & 0 \\
0 & 1 & 1 & 0 & 0 & 1 & 0 & 0 & 0 & 1 & 0 & 0 \\
0 & 0 & 1 & 0 & 0 & 1 & 1 & 0 & 0 & 0 & 0 & 0 \\
0 & 0 & 0 & 1 & 1 & 0 & 0 & 1 & 0 & 0 & 1 & 0 \\
0 & 0 & 0 & 0 & 1 & 0 & 0 & 1 & 0 & 0 & 0 & 0 \\
0 & 0 & 0 & 0 & 0 & 1 & 1 & 0 & 0 & 0 & 0 & 1 \\
0 & 1 & 0 & 0 & 0 & 0 & 0 & 0 & 0 & 1 & 0 & 0 \\
0 & 0 & 0 & 1 & 0 & 0 & 0 & 0 & 1 & 0 & 1 & 0 \\
0 & 0 & 0 & 0 & 0 & 1 & 0 & 0 & 0 & 1 & 0 & 1 \\
0 & 0 & 0 & 0 & 0 & 0 & 0 & 1 & 0 & 0 & 1 & 0 \\
\end{array}
$$

* In a general graph it is also allowable that there be several distinct links from a given vertex to another given vertex. We shall not consider this possibility here.

Now let us modify the flow chart and program of the preceding section in the following way. We no longer assume that there is a link from i to j for every pair i, j of vertices. The input to the computer will now include the matrix $W = (w_{ij})$ which is a modification of the matrix $T = (t_{ij})$. The modification is this: for a pair i, j for which there is no link from i to j, the number w_{ij} is arbitrarily given the value 0. For other pairs i, j, w_{ij} is taken equal to t_{ij}. Thus, the entries in W can be defined by the equations

$$w_{ij} = t_{ij} \quad \text{if} \quad c_{ij} = 1$$
$$\phantom{w_{ij}} = 0 \quad \text{if} \quad c_{ij} = 0 \tag{1}$$

In particular, $w_{ii} = 0$.

As before, the input to the computer also includes as initial guess $f_j^{(1)}$ $(j = 1, \cdots, N)$. Now when we use the recurrence equation

$$f_i^{(k+1)} = \min_j (t_{ij} + f_j^{(k)}) \tag{2}$$

we perform the minimization only for those j for which $w_{ij} \neq 0$, or equivalently* for which $c_{ij} \neq 0$. That is, the minimum is sought among those j for which there is a link from i to j.

The relation in (2) now becomes

$$f_i^{(k+1)} = \min_{w_{ij} \neq 0} (w_{ij} + f_j^{(k)}) \tag{3}$$

Exercises

1. Write a program using the method of this section and use it to compute Table 6.

2. The *complementary matrix* C' for a graph G is obtained by changing each 1 to 0 and each 0 to 1 in its connection matrix. This matrix C' can be considered to be the connection matrix of a different graph G'. For the graph in Fig. 8, find the complementary matrix and the complementary graph.

3. For the connection matrix C given above compute the matrix product $C^2 = C \times C$. Show that the entry in row i and column j of C^2 represents the number of distinct paths from vertex i to vertex j consisting of exactly two links. (Readers unfamiliar with the notion of matrix product should consult textbooks on matrices.)

4. Show that the conclusion of the preceding exercise is valid for an arbitrary connection matrix C. What is the significance of the elements in C^3, C^4, and so on?

---

* We make the reasonable assumption that if there is a link from i to j, then t_{ij} is positive, not zero.

5. Suppose that for some integer *m* all entries of the matrix C^m are zero. Show then that the graph contains no closed circuits. Is the converse true?

21. Fictitious Links

Another way in which to treat graphs for which the quantity t_{ij} does not exist for every pair of vertices is to imagine that there is a direct connection and to assign the time infinity (∞) to each such fictitious link. Every blank space in the array of times t_{ij} is thus replaced by ∞. This convention can change neither the value of the quickest time nor the quickest route from a vertex to the terminal vertex since a finite minimum cannot be attained for a *j* for which $t_{ij} = \infty$.

In the fundamental equations

$$f_i = \min_{j \neq i} (t_{ij} + f_j) \tag{1}$$

the index *j* can now be allowed to range over all integers 1 to *N*, excluding only *i*. Likewise in the equations which define the successive approximations, *j* ranges over all integers different from *i*.

On the other hand, if we are going to perform the required calculations on a digital computer, we cannot really assign $t_{ij} = \infty$ for fictitious links. No computer will accept the value "∞"; it is not in its vocabulary. In fact, such machines operate with a "finite field arithmetic," which means that they operate entirely with numbers no larger than a certain fixed largest one, perhaps 10^{30}, and no smaller than a fixed smallest one, say -10^{30}. For practical purposes this is quite adequate, but it does mean that the machine cannot routinely carry out the procedure based on Eq. (1).

To get around this difficulty, we can in practice assign to each fictitious link a time

$$t_{ij} = L \tag{2}$$

where *L* is a very large number, but one that is representable within the machine to be used. It will suffice to take *L* larger than the sum of all the times t_{ij} for nonfictitious links, since then no fictitious link can possibly appear as part of a quickest route. By way of illustration, if we apply this idea to the map of Fig. 3 of Chapter One and take $L = 1000$, the array of times will appear as in Table 8. We leave it to the reader to carry out the calculation of the quickest routes and to verify that there is no change in the final results.

By this device, we make it possible to use the computer program in § 19 for any routing problem. It is only necessary to store the full matrix t_{ij} in the machine, including a large entry *L* for each fictitious link. Thus, the use of fictitious links affords an alternative to the use of the connection matrix in dealing with the fundamental equations. Still another alternative will be presented in Chapter Three.

TABLE 8

To From	0	1	2	3	4	5	6	7	8
0	0	28	50	1000	1000	1000	1000	1000	1000
1	28	0	1000	46	1000	1000	1000	1000	1000
2	40	1000	0	27	52	1000	1000	1000	1000
3	1000	46	27	0	1000	47	1000	1000	1000
4	1000	1000	37	1000	0	16	103	1000	1000
5	1000	1000	1000	17	26	0	1000	62	1000
6	1000	1000	1000	1000	118	1000	0	47	1000
7	1000	1000	1000	1000	1000	82	47	0	43
8	1000	1000	1000	1000	1000	1000	1000	33	0

Exercises

1. Compute the quickest times f_i using the data in Table 8 and the initial approximation in (15.1).

2. Compute the quickest times f_i using the data in Table 8 and the initial approximation $f_i^{(1)} = 1000$ $(i = 0, 1, \cdots, 7)$, $f_8^{(1)} = 0$. For a fixed i, look at the numbers $f_i^{(1)}$, $f_i^{(2)}$, $f_i^{(3)}$, etc. At what point do they drop below 1000?

22. Paths With a Given Number of Junctions

Certain choices of the initial values $f_i^{(1)}$ produce numbers $f_i^{(k)}$ with a simple physical significance. In this section we shall discuss one such choice, that in which $f_N^{(1)} = 0$ and every other $f_i^{(1)}$ is a "large number." For theoretical purposes, it is useful to interpret "large number" here as meaning $+\infty$, but for practical purposes (and especially for machine calculation), any number which exceeds the magnitude of every possible sum of distinct t_{ij}'s will do. See the discussion in the preceding section.

Let us set

$$f_i^{(1)} = +\infty \qquad i = 1, 2, \cdots, N - 1$$
$$f_N^{(1)} = 0 \tag{1}$$

and see if we can discern the physical meaning of the subsequent values $f_i^{(k)}$.

It is convenient if we first consider the case in which there is a link joining every pair of vertices, or in other words, t_{ij} is defined and finite for every i and j.

We shall illustrate our arguments by reference to the map in Fig. 12, with an array of times as in Table 9.

TABLE 9

From To	1	2	3	4
1	0	5	15	40
2	5	0	5	35
3	15	5	0	20
4	40	35	20	0

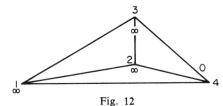

Fig. 12

Here we suppose the desired destination is vertex 4, that is, $N = 4$. In Fig. 12, the number $f_i^{(1)}$ has been written next to vertex i for each i.

Now we compute $f_i^{(2)}$, using the usual recurrence relations

$$f_i^{(2)} = \min_{i \neq j} (t_{ij} + f_j^{(1)}) \qquad i = 1, 2, 3$$
$$f_4^{(2)} = 0 \tag{2}$$

The computation can be performed in the same way as before by making a strip with the numbers $f_i^{(1)}$ and placing it over Table 9. It turns out that $f_1^{(2)} = 40$, $f_2^{(2)} = 35$, and $f_3^{(2)} = 20$. We see that $f_i^{(2)}$ represents simply the time required for the one-link path directly from i to 4. In fact it is clear from Eq. (2) that $t_{ij} + f_j^{(1)}$ is infinite except when $j = 4$ and is t_{i4} when $j = 4$. Therefore

$$f_i^{(2)} = t_{i4} \qquad i = 1, 2, 3 \tag{3}$$

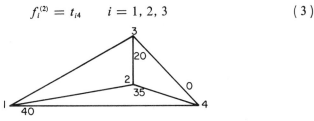

Fig. 13

The numbers $f_i^{(2)}$ are shown next to their respective vertices in Fig. 13. Now let us calculate the numbers $f_i^{(3)}$, using

$$f_i^{(3)} = \min_{j \neq i} (t_{ij} + f_j^{(2)}) \qquad i = 1, 2, 3$$
$$f_4^{(3)} = 0 \qquad\qquad\qquad\qquad (4)$$

For example,

$$f_1^{(3)} = \min (t_{12} + f_2^{(2)}, \ t_{13} + f_3^{(2)}, \ t_{14} + f_4^{(2)}) .$$

Since $f_j^{(2)} = t_{j4}$ for $j = 1, 2, 3$ and $f_4^{(2)} = 0$, we have

$$f_1^{(3)} = \min (t_{12} + t_{24}, \ t_{13} + t_{34}, \ t_{14} + 0)$$
$$= \min (5 + 35, \ 15 + 20, \ 40) = 35$$

We observe that $f_1^{(3)}$ is the least time for a path from 1 to 4 which has either one or two links. Similarly we get $f_2^{(3)} = 25$; this represents the time for the path from vertex 2 to vertex 3 to vertex 4, and is the least time for a path from 2 to 4 with one or two links. Finally, $f_3^{(3)} = 20$, and this is the least time for a path with one or two links from 3 to 4 (in this case the one-link path is the best). The numbers $f_i^{(3)}$ are shown in Fig. 14.

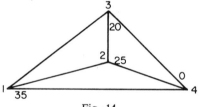

Fig. 14

In the same way, we compute $f_i^{(4)}$, obtaining $f_1^{(4)} = t_{12} + f_2^{(3)} = 30$, $f_2^{(4)} = t_{23} + f_3^{(3)} = 25$ and $f_3^{(4)} = t_{34} = 20$. It is not hard to see that these represent the least times for paths to vertex four with one, two, or three links.

If now we compute $f_i^{(5)}$ from the recurrence relation, we find that $f_i^{(5)} = f_i^{(4)}$ $(i = 1, 2, 3, 4)$. It is easy to see why these numbers are equal. Indeed, by extending the argument above we can show that $f_i^{(5)}$ represents the minimal time for a path from i to 4 with no more than four links. However, it is clear from the given map that there are no four-link paths except those with circuits. Therefore the minimal time is in fact achieved by a path with at most three links, and so the time $f_i^{(5)}$ must be the same as the time $f_i^{(4)}$. Observe that the physical interpretation of the numbers $f_i^{(k)}$ has enabled us to establish the convergence in a finite number of steps of the successive approximation process. In Chapter Four, we shall use the same type of argument to establish convergence for an arbitrary map and an arbitrary choice of the numbers $f_i^{(1)}$.

Now suppose we consider *any* map for which there is a link joining every pair of vertices. Define

$$f_i^{(1)} = \infty \qquad i = 1, 2, \cdots, N - 1$$
$$f_N^{(1)} = 0$$

and define $f_i^{(k)}$ by the general recurrence relation

$$f_i^{(k)} = \min_{j \neq i} (t_{ij} + f_j^{(k-1)}) \qquad i = 1, 2, \cdots, N - 1$$
$$f_N^{(k)} = 0 \tag{5}$$

Then we assert that $f_i^{(k)}$ *is the least time for a path with at most $k - 1$ links from vertex i to vertex N* $(i = 1, 2, \cdots, N - 1; k = 2, 3, \cdots)$.

The general proof is by mathematical induction. For $k = 2$, the assertion is clearly true since $f_i^{(2)} = t_{iN}$. Assume that the assertion is correct for a certain integer k. Consider any fixed vertex $i (i \neq N)$, and the class C_k of all paths from i to N with at most k links. One of these paths goes directly to N and the associated time is

$$t_{iN} = t_{iN} + f_N^{(k)}$$

The other paths go from i to j to $N (i \neq j \neq N)$ for some j. With i and j fixed, the associated time is least if the path from j to N is the optimal one with $k - 1$ links, which by the assumption takes time $f_j^{(k)}$. Therefore, the optimal path of at most k links which goes first to j takes time $t_{ij} + f_j^{(k)}$. It follows that the path of least time in the class C_k requires time

$$\min (t_{ij} + f_j^{(k)}) \qquad (j = 1, \cdots, N; \ j \neq i)$$

and this is $f_i^{(k+1)}$ by definition. Thus $f_i^{(k+1)}$ does represent the least time for a path with at most k links, as we had to show. This completes the proof.

If there are pairs of vertices in the map which are not joined by links, the interpretation of $f_i^{(k)}$ has to be modified. We leave it as an exercise to show that in this case *if the initial approximation $f_i^{(1)}$ is defined by (1) then $f_i^{(k)}$ is the least time for a path with at most $k - 1$ links from vertex i to vertex $N (i = 1, 2, \cdots, N - 1)$ if any such path exists, and $f_i^{(k)} = \infty$ if there is no path with at most $k - 1$ links from i to N.*

There is a slightly different way in which we can establish the above interpretation of the numbers $f_i^{(k)}$. Suppose we define $f_i^{(k)}$ by (1) and (5), but let $g_i^{(k)}$ be *defined* as the least time to reach N from i on a path with at most $k - 1$ links. On this minimal path, one must travel first from i to an accessible vertex j, and then follow the minimal path with at most $k - 2$ links from j to N. Thus we can show that

$$g_i^{(k)} = \min_{j \neq i} (t_{ij} + g_j^{(k-1)}) \qquad i = 1, \cdots, N - 1$$
$$g_N^{(k)} = 0 \tag{6}$$

But now each $g_i^{(k)}$ satisfies the same equation as the corresponding $f_i^{(k)}$. Moreover, $g_i^{(2)} = t_{iN} = f_i^{(2)}$, (interpreted as ∞ if there is no link from i to N). Since the $f_i^{(k)}$ and $g_i^{(k)}$ are the same for $k = 2$, and satisfy the same equations for each k, it follows inductively that $f_i^{(k)} = g_i^{(k)}$ for all i and k. In other words, we rely on the fact that the equations in (6) define a *unique* set of numbers. This is the advantage of a recurrence relation, as opposed, say, to the equation of (1.1).

Exercises

1. Consider the map in Fig. 15 with array of times as in Table 10.

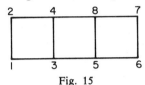

Fig. 15

TABLE 10

To From	1	2	3	4	5	6	7	8
1	0	6	5					
2	6	0		2				
3	5		0	2	1			
4		2	2	0				5
5		1			0	2		3
6					2	0	2	
7						2	0	8
8							8	0

Using $f_i^{(1)}$ as defined by (1), find $f_i^{(2)}$, $f_i^{(3)}$, and $f_i^{(4)}$, and verify that these numbers represent the shortest paths with at most 1, 2, and 3 links, respectively, if such paths exist, and are ∞ if no such paths exist.

2. Consider the equations for successive approximations formulated in Exercise 3 of §18. If the initial policy is $f_1^{(1)} = 0$, $f_i^{(1)} = \infty$ ($i = 2$, $3, \cdots, N$), what is the physical interpretation of the numbers $f_i^{(k)}$?

23. A Labelling Algorithm

For a small map such as in Fig. 12, the iteration process defined by (22.1) and (22.5) can be performed very easily by hand by means of the following algorithm, which we shall call a "labelling algorithm."

Step 1. Write a zero beside the terminal vertex, N, and ∞ beside each of the others.

Step 2. For each vertex i, other than N, compute the sum of t_{ij} and the label on vertex j, for each $j \neq i$. Compute the minimum m_i of these sums and record m_i.

Step 3. After step 2 has been executed for every $i \neq N$, change the label on each i to the minimum m_i computed in step 2.

Step 4. Return to steps 2 and 3 again and compute new labels. Repeat steps 2 and 3 until the labels no longer change.

For the example of Section 22, Fig. 12 is the result of step 1 of this algorithm. After doing steps 2 and 3 once, we get Fig. 13. Doing steps 2 and 3 again, we get Fig. 14 and then further repetition of steps 2 and 3 does not alter the labels.

We see that after the first use of steps 2 and 3 the labels are the numbers $f_i^{(2)}$, the minimal times for paths to N with at most one link, after another use of steps 2 and 3 the labels are the numbers $f_i^{(3)}$, the minimal times for paths to N with at most two links, and so on.

Exercises

1. Carry through the labelling algorithm for the map in Fig. 15.

2. For the map of Fig. 11 of Chapter One (see the array of times in Table 5 of this chapter), find the least time in which one can go from vertex i to vertex 11 on a path with at most four steps, for $i = 1, 2, 3, 4$.

24. Working Back From the End

If the reader has carried through the above algorithm for Fig. 15, he has no doubt noticed that a number of steps are performed which do not produce any improved information. For example, the first time step 2 is performed for vertex 1, the instructions call for us to compute $t_{12} + \infty = \infty$ and $t_{13} + \infty = \infty$, and to record $m_1 = \infty$. For a map of such a small number of vertices, these unprofitable calculations can be eliminated "by inspection." However, with a view to maps of larger size, and to mechanization of our procedures for a digital computer, we wish to outline an algorithm which eliminates these unnecessary calculations by the simple expedient of not computing an approximation until a finite value can be supplied.

The preliminary stage of this algorithm consists in dividing the vertices of the given map into sets S_1, S_2, S_3, etc., in the following way. In the set S_1 we place the single vertex N. In the set S_2 we place N and all vertices from which vertex N can be reached in one step (that is, on a one-link path). In S_3 we place all vertices in S_2 and all vertices from which a vertex in S_2 can be reached in one step. In general, we place in S_k all vertices (including those already in S_{k-1}) from which a vertex in S_{k-1} can be reached in one step.

For Fig. 15

$$
\begin{aligned}
S_1 &= \{\, 8 \,\} \\
S_2 &= \{\, 4,\ 5,\ 7,\ 8 \,\} \\
S_3 &= \{\, 2,\ 3,\ 4,\ 5,\ 6,\ 7,\ 8 \,\} \\
S_4 &= \{\, 1,\ 2,\ 3,\ 4,\ 5,\ 6,\ 7,\ 8 \,\}
\end{aligned}
\tag{1}
$$

Here we have used the notation $\{a, b, c, \cdots\}$ to denote the set whose elements are $a, b, c, \cdots$, as is customary in the theory of sets.

Once the sets S_k have been determined, we begin a successive approximations scheme by setting

$$
f_N^{(1)} = 0 \tag{2}
$$

Next, for each vertex i in S_2 we compute

$$
f_i^{(2)} = \min_j (t_{ij} + f_j^{(1)}) \qquad i \in S_2
$$

where the minimum is taken over all j in S_1. Since there is only one point in S_1, namely $j = N$, this gives

$$
f_i^{(2)} = t_{iN} \qquad i \in S_2 \tag{3}
$$

No value is computed for $f_i^{(2)}$ if i is not in S_2.

At the next step we compute

$$
f_i^{(3)} = \min (t_{ij} + f_j^{(2)}) \qquad i \in S_3 \tag{4}
$$

Here the minimum is taken over all j for which $f_j^{(2)}$ and t_{ij} are defined. In general, the recurrence relation is

$$
f_i^{(k)} = \min (t_{ij} + f_j^{(k-1)}) \qquad i \in S_k \tag{5}
$$

The minimum is taken over j in S_{k-1} for which t_{ij} is defined.

It is not hard to see that by this procedure we obtain the same values for $f_i^{(k)}$ that we obtained by our former method, except that we do not compute with any "infinite" values. It is still true that $f_i^{(k)}$ is the time for the quickest path from i to N with at most $(k - 1)$ links. A value is computed for $f_i^{(k)}$ if and only if a path from i to N exists having no more than $(k - 1)$ links.

If we perform this iteration for Fig. 15, we obtain

$$
f_8^{(1)} = 0
$$

then

$$f_4^{(2)} = 5 \qquad f_5^{(2)} = 3 \qquad f_7^{(2)} = 8 \qquad (6)$$

At the next step, we get

$$f_4^{(3)} = 5 \qquad f_5^{(3)} = 3 \qquad f_7^{(3)} = 8$$
$$f_2^{(3)} = 7 \qquad f_3^{(3)} = 4 \qquad f_6^{(3)} = 5 \qquad (7)$$

Notice that values for vertices 4, 5, and 7 are recomputed at this step. For Fig. 15, this cannot change the values since no paths with 2 links can reach vertex 8 from vertices 4, 5, or 7.

At the next step, we get

$$f_4^{(4)} = 5 \qquad f_5^{(4)} = 3 \qquad f_7^{(4)} = 7$$
$$f_2^{(4)} = 7 \qquad f_3^{(4)} = 4 \qquad f_6^{(4)} = 5 \qquad (8)$$
$$f_1^{(4)} = 9$$

Note that the recomputed value for f_7 is an improvement brought about because the path 7 to 6 to 5 to 8 is quicker than the link 7 to 8.

Since there is no change if we perform one more iteration, the values in (8) are the optimal values.

As the reader can see, the above method can be carried out by successive labellings of the vertices on a map. First, vertices in S_1 and S_2 are labelled. Next, vertices in S_3 are labelled (the labels on vertices in S_2 being perhaps changed), and so forth. The labelling is thus done in layers, *working back from the destination*. At each successive step, labels already obtained are subject to revision.

The iteration scheme here is clearly much more efficient than the previous method in which the initial values (22.1) are used, if it is true that each vertex is connected to only a few others, as is normally the case for a street or road network. Of course, if every vertex is connected to every other, the two methods are identical. Note that the determination of the sets S_k given the connection matrix or a list of vertices preceding each given vertex, itself requires considerable calculation.

On the other hand, when we use the algorithm defined by (2) and (5), we give up the possibility of adopting an initial policy which may appear to be close-to-optimal. For example, suppose that for the map in Fig. 15, we choose $f_i^{(1)}$ to be the time for the path from i to 8 which goes straight across until it reaches 8 or 5, and then up from 5 to 8 if necessary. This gives

$$f_1^{(1)} = 9 \qquad f_2^{(1)} = 7 \qquad f_3^{(1)} = 4 \qquad f_4^{(1)} = 5$$
$$f_5^{(1)} = 3 \qquad f_6^{(1)} = 5 \qquad f_7^{(1)} = 8 \qquad f_8^{(1)} = 0 \qquad (9)$$

Now iteration as in equations (18.2) gives us values $f_i^{(2)} = f_i^{(3)}$ which are the optimal ones. Thus, once the initial values are selected, we are sure we have the true values after two iterations, or in other words after 14 calculations of a minimum. By the other method, 6 such calculations are required to get (7), 7 to get (8), and 7 more to show that the values in (8) do not change on the next iteration.

In Chapter Three we shall have more to say about the relative efficiency of these methods, as well as others. The point we want to make now is that there are a number of variations of our fundamental method. In any given case, one must adopt a technique suited to the information at hand and the computing facilities available.

Exercises

1. Write a program to generate the sets S_1, S_2, etc., given the connection matrix.

2. Describe a similar labelling algorithm for finding the minimal path from a fixed starting vertex, 1, to each other vertex in a graph.

Miscellaneous Exercises

1. One of the most efficient labelling algorithms was published by E. W. Dijkstra in 1959. This algorithm finds the quickest route from a starting vertex to every other vertex. At each stage some nodes will have permanent labels and others will have temporary labels. The algorithm is as follows:

    Step 1.   The starting vertex is given the permanent label 0 and each other vertex is given the temporary label ∞.

    Step 2.   (a) A new temporary label is assigned to each vertex j which is not yet permanently labeled. The rule for this is

    $$j\text{'s new label} = \min (\text{old label}, t_{vj} + \text{label on } v),$$

    where v denotes the vertex which most recently received a permanent label.

    (b) The smallest of the new temporary labels is found, and this becomes a permanent label.

    Step 2 is repeated until all vertices have permanent labels.

    Carry out this algorithm for the map in Fig. 15. Whenever a new permanent label is assigned to a vertex j, record the vertex from which j is reached. Show how to construct the quickest routes from this record. See

    E. W. Dijkstra, "A Note on Two Problems in Connection with Graphs," *Numerische Math.* **1** (1959) 269–271.

    S. E. Dreyfus, *An Appraisal of Some Shortest Path Algorithms*, RAND Corp., RM-5433-PR, Santa Monica, California, October, 1967.)

2. Show that to determine all permanent labels is Dijkstra's algorithm for a graph with N vertices requires $N(N-1)/2$ additions and $N(N-1)$ comparisons. (Additional steps are needed to keep the record of the vertex from which each vertex is reached.)

3. Write a flow chart and computer program for Dijkstra's algorithm.

4. Consider a graph with N vertices and let P be the quickest path from vertex 1 to vertex N. A *deviation* from P is defined to be a path that coincides with the quickest path up to some node j on the path (j may be 1 or N), then goes directly to some node k not the *next* node on P, and finally proceeds from k to N via the quickest route from k to N.

   (a) In Figure 15, find the quickest path from each node j to 8.
   (b) List all deviations from the quickest path from 1 to 8.
   (c) Which deviation takes the least time? Compare it with the second quickest path from 1 to 8. See

W. Hoffman and R. Pavley, "A Method for the Solution of the *N*th Best Path Problem," *J. Assoc. Comp. Mach.* **6** (1959) 506–514.

R. Bellman and R. Kalaba, "On *k*th Best Policies," *J. SIAM* **8** (1960) 582–588.

S. E. Dreyfus, *op. cit.*

5. Give a simple example of a graph in which the second quickest path from 1 to N passes through N twice; hence, contains a circuit.

6. A solution of the shortest-route problem for simple cases can be found with the aid of a string model in the following way. In this model, knots represent cities and the length of string between two knots is proportional to the travel time t_{ij} between the corresponding cities. (It is assumed here that $t_{ij} = t_{ji}$.) After the model is finished, take the knot representing the initial vertex in one hand and the knot representing the terminal vertex in the other hand and pull them apart. If the model becomes entangled, untie and re-tie knots until the entanglements disappear. Eventually one or more paths will stretch tight. These are then the quickest paths. Build such a string model for one of the examples discussed in the text and verify that the correct answer is obtained. See

George J. Minty, "A Comment on the Shortest-Route Problem," *Oper. Research* **5** (1957) 724.

7. The method used in the preceding exercise is a primitive example of the use of an *analog computer*, which is simply a device which replaces a given problem by an analogous physical system. The solution to the problem is then obtained by an experiment using the physical system. Can you think of another analog device for solving the shortest-

route problem which uses gas-discharge tubes (such as neon lamps) which conduct current only above a certain critical voltage? This method, due to F. Bock and S. Cameron, is discussed by

R. M. Pearl, P. H. Randolph, and T. E. Bartlett, "The Shortest-Route Problem," *Oper. Research* **8** (1960) 866–868.

See also

H. Rapoport and P. Abramson, "An Analog Computer for Finding an Optimum Route Through a Communications Network," *IRE Transactions on Communications Systems* CS-7 (1959) 37–42.

V. L. Klee, "A 'String Algorithm' for Shortest Path in Directed Networks," *Oper. Research* **12** (1964) 428–432.

8. A problem closely related to the shortest route problem is this. In a network (graph) with directed edges, let p_{ij} be the probability that the edge from vertex i to vertex j will be operative. Then for each path in the network the probability that the path is operative is the product of the probabilities for its various edges. Construct an algorithm for finding the path with greatest probability of being operative between two specified vertices. See

O. Wing, "Algorithms to Find the Most Reliable Path in a Network," *I.R.E. Transactions on Circuit Theory*, CT-8 (1961) 78–79.

G. R. Shorack, "Algorithms and Analog Computers for the Most Reliable Route Through a Network," *Oper. Research* **12** (1964) 632–633.

9. What are the fundamental equations for the unknowns p_i in the previous exercise if p_i denotes the largest possible probability of being operative for a path from vertex i to the terminal vertex N? (Take $P_N = 1$.)

10. Given a graph, let c_{ij} denote the "capacity" of the directed edge from vertex i to vertex j. The capacity of a path is defined to be the smallest of the capacities of the edges comprising the path. The maximum capacity route between two vertices is the path between these vertices with the largest capacity. Devise and compare algorithms for determination of the maximum capacity route. See

M. Pollack, "The Maximum Capacity Route Through a Network," *Oper. Research* **8** (1960) 733–736; Errata, **9** (1961) 133.

11. Devise a labelling algorithm for the determination of a critical path as discussed in § 19 of Chapter One.

12. (The Golden Spike) When the first transcontinental railway was constructed in the U. S., the line was built from both ends at the same time. The Central Pacific line was laid eastward, beginning at San Francisco, and the Union Pacific line was laid westward, beginning at

Omaha, Nebraska. The laying of the rails proceeded at great speed until the two tracks were ceremoniously joined at Promontory Point, Utah, on May 10, 1869, by the driving of a golden spike. A problem which might have been considered by the railroad executives and the United States Congress is suggested by this historical event, namely, given the two starting points, the alternative routes available for crossing the western mountains, and given the times required to lay a mile of track under the various circumstances to be encountered (on the plains, in the foothills, in the mountains, etc.), what route should be selected in order that the tracks be joined in minimum time?

13. An analogous problem in a simpler setting is as follows: A relay team consists of two racers. One starts at vertex i and the other at a vertex j in a network. They choose a vertex k (possibly i or j itself) and drive to it. The time recorded for the team is the larger of the driving times of its two members. Now given i and j, how shall they choose k, and what is the minimum time the team can achieve?

14. Suppose that in the relay race of Problem 13 the time for a team is the sum of the times for its two members. Now explain how to find k and the team time, given the starting vertices i and j.

15. (The Life-Saving Drug) A patient has been struck by a sudden illness, and can be saved only by prompt injection of a new drug, a product of scientific labor called Miracure. If the drug reaches either one of two known vital organs within the body in a short enough time, a full cure is assured; unfortunately, if the drug is delayed more than a few minutes, it is completely ineffective. Of the several places where the drug can be injected—the arm, calf, hip, and so on—which one should be chosen?

   Discussion. Let us suppose that the drug has entered the blood stream, and is being carried through the circulatory system. Now and then a branching of arteries or veins occurs, or the blood passes through an organ. As a first rough model of what happens, we can think of the circulatory system as forming a network of passageways through which blood and the injected drug may pass, much as a road system is a network of passageways for cars. We can imagine that wherever blood vessels branch, part of the injected drug will flow in each possible direction. Consequently a portion of the drug will reach one of the vital organs in a time equal to the time for a quickest route from the injection site to the organ. Thus the problem reduces to the following form: given a network and two of its vertices, say i and j, find the vertex from a given set $v_1, \cdots, v_p$ of vertices from which one can reach either i or j in least time. Outline a method for finding the best injection site in this situation.

16. Suppose that in the above problem a cure cannot be achieved until Miracure reaches both vital organs. Outline a method for finding the best injection site in this situation. (These problems were contributed by D. L. Bentley.) See also

A. Munck, "Symbolic Representation of Metabolic Systems," *Mathematical Biosciences* (forthcoming).

17. The problem to be solved is that of maximizing the profit gained by a commercial airline from a single flight from New York City to Los Angeles. This particular problem is not entirely realistic since the goal of an airline is not merely profit maximization of one flight, but its maximization over an entire schedule of flights while providing the acceptable customer service. The solution should point out what route should be followed by a large number of intermittent flights where the concern for customer service has been alleviated by the scheduling of other flights.

The airplane with which this problem will be concerned will be allowed to land and take on passengers at any of $N - 2$ cities on its flight from New York City, city 1, to Los Angeles, city N. The revenue gained from the flight will be equal to the number of passengers boarding the plane multiplied by a cost to the passenger that is proportional to the direct distance to Los Angeles from the city at which he boards. The number of people boarding at each city will be proportional to the population of the city. There will be no limit on the capacity of the airplane.

The cost of the flight will be represented by three different quantities. The first cost to be considered will be proportional to the total distance covered by the plane. The second will be a fixed cost of landing at each city. The last will be what might be called a "complexity cost." It will be expressed in terms of stops and the input-output aspects of passengers using terminals. How does one plan a flight schedule?

18. In many situations it is important not only to determine optimal routing, but also to determine ways of improving the existing routing. At very least, we would like to do better than what has been done before. Let $j(i)$ be an existing policy, i.e., a rule which tells us what vertex to proceed to next from i. Let g_i be the time consumed going from i to N using this policy. Show that g_i satisfies the equation $g_i = t_{ij} + g_j$, where $j \equiv j(i)$.

19. Hence, show that g_i can be calculated using the expression

$$g_i = t_{ij_1} + t_{j_1 j_2} + t_{j_2 j_3} + \cdots,$$

where $j_1 = j(i)$, $j_2 = j(j_1)$, $j_3 = j(j_2)$, $\cdots$.

20. Suppose that we calculate a new policy, $k(i)$, by choosing $k(i)$ to be the value of j which minimizes $t_{ij} + g_j$. Let h_i be the times calculated using this new policy $k(i)$. Can we assert that $h_i \leq g_i$, $i = 1, 2, \cdots$, $N - 1$? What is the meaning of this procedure? This is an application of the technique of "approximation in policy space." For important applications of this general idea, in the form of "policy improvement methods," see

R. Howard, *Dynamic Programming and Markov Processes*, Wiley, New York, 1960.

Bibliography and Comment

One of the earliest papers to give a clear statement of the shortest route problem was

A. Shimbel, "Structure in Communication Nets," *Proceedings of Symposium on Information Networks*, Polytechnic Institute of Brooklyn, April 12-14, 1954.

Shimbel gave a solution based on operations with what we have called the matrix of the graph. A labelling algorithm essentially equivalent to the one we have presented in § 24 was given by E. F. Moore, and is perhaps the earliest algorithm of this type; see

E. F. Moore, "The Shortest Path Through a Maze," *Proceedings Internat. Symp. on the Theory of Switching*, Part II, April 2-5, 1957. *Annals of the Computation Laboratory of Harvard University* **30** (1959) 285-292, Harvard U. Press, Cambridge, Mass.

As we noted in the Bibliography and Comments for Chapter One, the dynamic programming approach to the routing problem first appeared in 1958.

A large number of interesting algorithms for the routing problem have subsequently been published by various authors. For a critical discussion of many of these, with a comparison of efficiencies, see the paper by S. Dreyfus, "An Appraisal of Some Shortest Path Algorithms," already cited in the Bibliography and Comments on Chapter One. This paper has an extensive bibliography and mentions a number of important extensions of the basic problem. Other bibliographical papers are

S. W. Stairs, *Bibliography of the Shortest Route Problem*, Transport Network Theory Unit, London School of Economics, Report LSE-TNT-6, revised June, 1966.

M. Pollack and W. Wiebenson, "Solution of the Shortest-Route Problem—A Review," *Oper. Research* **8** (1960) 224-230.

See also

R. D. Wollmer, "Sensitivity Analysis of Maximum Flow and Shortest Route Networks," *Management Science* **14** (1968) 551-564.

B. A. Farbey, A. H. Land, J. D. Murchland, "The Cascade Algorithm for Finding all Shortest Distances in a Directed Graph," *Management Science* **14** (1967) 19-28.

See the Bibliography and Comments in Chapter Three for additional bibliographical information, including references to papers containing computer programs.

For a discussion of the Newton-Raphson method and extensions, see

R. Bellman and R. Kalaba, *Quasilinearization and Nonlinear Boundary-Value Problem*, American Elsevier, New York, 1965.

The reader interested in learning about the theory of algorithms can refer to the book

B. A. Trakhtenbrot, *Algorithms and Automatic Computing Machines*, D. C. Heath, Boston, 1963.

It is extremely interesting to observe the repercussions of a technological device, the digital computer, upon some of the most abstract and recondite areas of mathematics.

Chapter Three

FROM CHICAGO TO THE GRAND CANYON BY CAR AND COMPUTER:
Difficulties Associated With Large Maps

1. Introduction

In this chapter we wish to discuss some new types of problems arising from the consideration of routing problems associated with large maps, which is to say those containing a large number of vertices. To illustrate these matters we shall begin by examining a specific problem, that of determining the route of shortest time between Chicago and the Grand Canyon.

Following this, we will describe a modification of the original method which can be used to bypass some of the bookkeeping aspects of carefully selected classes of optimal routing problems. This technique is extremely useful in connection with the calculation of optimal trajectories and, more generally, with the control and scheduling of the activities of large economic or engineering systems.

In some cases, the specialized geometrical structure of the network can be used with considerable success to overcome the dimensionality difficulty, as our example will demonstrate.

By the term "dimensionality difficulty" we mean a complex of problems, namely those that arise when the number of vertices in the map becomes considerable. There is a philosophical dictum to the effect that a change of quantity becomes a change of quality. This is nowhere better illustrated than in the use of the computer to obtain numerical solutions.

All of the questions we discuss are associated with the study of large systems, one of the new and exciting domains of modern mathematics.

101

Our aim is to give the reader a brief inkling of the new types of problems that arise and of the kinds of methods that can be used to overcome and circumvent obstacles to feasible solution without becoming too deeply embroiled or diverted from our principal goal of introduction and exposition.

2. How to Plan a Vacation

The maps which we have heretofore used for illustration have been maps of city streets which are laid out on an essentially rectangular grid. However, our methods apply equally well to road networks of any description. As an illustration of this, suppose that a family in Chicago is planning a vacation trip to the Grand Canyon. Since only a limited time is available, they wish to choose a route from Chicago to the Grand Canyon which will require the least driving time.

In order to solve this problem, we can begin by consulting a road map of the United States and attempting to pick out the quickest route from among a rather sizeable number of possibilities. This choice may be based on a map which shows estimated driving times as well as mileages between cities. For example, we can use a Transcontinental Mileage and Driving Time Map from the Rand McNally Road Atlas, 40th Edition. From this map, the time for any particular route from Chicago to the Grand Canyon can be computed by adding the city-to-city times. However, it is clear that the number of possible routes is quite large. Certainly any painstaking enumeration will be both tedious and time-consuming.

The alternative method which we have previously applied to city maps can readily be applied here. The first step, as before, is to number the road intersections. As a practical matter, it is clear that the quickest route cannot deviate too far from a direct line from Chicago to the Grand Canyon. Consequently, we need not consider too large a map. In Fig. 1,

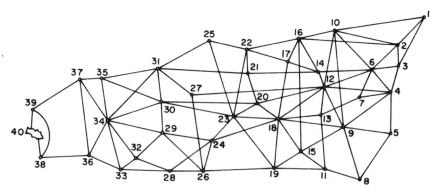

Fig. 1

TABLE 1

KEY TO MAP

1	Chicago	21	Phillipsburg, Kansas
2	Peoria, Illinois	22	Grand Island, Nebraska
3	Springfield, Illinois	23	Dodge City, Kansas
4	St. Louis, Missouri	24	Guymon, Oklahoma
5	Poplar Bluff, Missouri	25	North Platte, Nebraska
6	Hannibal, Missouri	26	Amarillo, Texas
7	Jefferson City, Missouri	27	Kit Carson, Colorado
8	Little Rock, Arkansas	28	Tucumcari, New Mexico
9	Springfield, Missouri	29	Raton, New Mexico
10	Des Moines, Iowa	30	Pueblo, Colorado
11	Ft. Smith, Arkansas	31	Denver, Colorado
12	Kansas City	32	Santa Fe, New Mexico
13	Ft. Scott, Kansas	33	Albuquerque, New Mexico
14	St. Joseph, Missouri	34	Durango, Colorado
15	Tulsa, Oklahoma	35	Grand Junction, Colorado
16	Omaha, Nebraska	36	Gallup, New Mexico
17	Beatrice, Nebraska	37	Green River, Utah
18	Wichita, Kansas	38	Flagstaff, Arizona
19	Oklahoma City, Oklahoma	39	Mt. Carmel Junction, Utah
20	Great Bend, Kansas	40	Grand Canyon

we reproduce a portion of the map which will evidently contain the quiekest path. The city names have been replaced by the numbers 1–40, from East to West, as indicated in Table 1, and the driving times have been given in minutes, in Table 2. Road intersections which do not occur at a labelled city have been disregarded. This means that changes of route can occur only at one of the forty labelled cities, since the Rand McNally Atlas gives driving times only between the labelled cities. If driving times between all road intersections were available, a more realistic solution could be found, at the expense of adding 36 additional vertices.

The fundamental equations

$$f_i = \min_{j \neq i} (t_{ij} + f_j) \qquad (1)$$

are valid, where f_i is, as above, the least time in which vertex 40 (Grand Canyon) can be reached from the vertex i, $i = 1, 2, \cdots, 39$. The method of successive approximations, as described by the equations

$$f_i^{(k)} = \min_{j \neq i} (t_{ij} + f_j^{(k-1)})$$
$$f_N^{(k)} = 0 \qquad (2)$$

TABLE 2

From \ To	1	2	3	4	5	6	7	8	9	10	11	12	13	14	15	16	17	18	19	20
1	0	220	255							460										
2	220	0	105			240				360						510				
3	255	105	0	130		165														
4			130	0	240	180	195		315			340								
5				240	0			300	310											
6		240	165	180		0	165			360		285		285						
7				195		165	0		225				255							
8					300			0	330		225									
9				315	310		225	330	0	505		255			220			360		
10	460	360				360			505	0	280				200					
11								225			0		330		210				270	
12				340		285			255	280		0	150	70				200		270
13							255				330	150	0		195			210		
14						285						70		0		255	195			
15									220		210		195		0	590		230	105	
16		510							220					255	590	0	110			
17														195		110	0	275		
18									360			200	210		230		275	0	195	170
19											270				105			195	0	
20												270						170		0
21														300						165
22															190	170				
23																210	360	120		
24																345				
25																				
26																		345		
27												285								
28																				
29																				
30																				465
31																				
32																				
33																				
34																				
35																				
36																				
37																				
38																				
39																				
40																				

TABLE 2 (*Continued*)

21	22	23	24	25	26	27	28	29	30	31	32	33	34	35	36	37	38	39	40	From / To
																				1
																				2
																				3
																				4
																				5
																				6
																				7
																				8
																				9
																				10
																				11
						585														12
																				13
300																				14
																				15
	190																			16
	170																			17
		210	345																	18
		360			345															19
165	120							465												20
0	160									390										21
160	0	375		190																22
	375	0	150	340					360											23
		150	0		150		260	255												24
	190	340		0						360										25
			150		0	360	175	285												26
					360	0				210										27
			260		175		0				260	230								28
			255		285			0	150		255		435							29
		360							0	180			435	480						30
390					360	210			180	0			580	420						31
							260	255			0	75	330							32
							230				75	0	300		185					33
								435	435	580	330	300	0	300	260	320				34
									480	420			300	0		180				35
												185	260		0	450	260			36
													320	180	450	0		480		37
																260	0	330	110	38
																480	330	0	150	39
																	110	150	0	40

will again be used to carry out the determination of the f_i. We shall attempt to choose an initial approximation adroitly.

First we shall carry out the calculations following the procedure outlined in the preceding chapter and then we will discuss some modifications which speed up the calculations.

3. A Particular Initial Approximation

A plausible choice of an initial approximation for the Grand Canyon problem is the following: $f_i^{(1)}$ is the time required for the route from i to N in which we always drive to the most westerly point accessible from i.

The first step is to determine the most westerly vertex which can be reached from a particular vertex i. In this case, this is the one with the highest number.* The path from any vertex to the Grand Canyon following this policy can then easily be read off and the travel time computed. In Table 3 below, the calculation has been begun, but has been left for the reader to complete.

After the calculation of $f_i^{(1)}$ for $i = 1, 2, 3, \cdots$, has been completed, the calculation of $f_i^{(k)}$ for $k \geq 2$ is carried out using the recurrence relation of (2.2), and a computer program.

If the calculations are also carried out by hand, which we recommend, it is desirable to work from the bottom of the table up. For example, after $f_{16}^{(1)}$ has been calculated, it is easy to compute $f_2^{(1)} = t_{2,16} + f_{16}^{(1)} = 510 + f_{(16)}^{(1)}$.

According to our calculations, the minimum time is 2290 minutes and the optimal route is [1, 3, 6, 12, 20, 23, 24, 28, 33, 36, 38, 40]. We leave it as an exericise for the reader to confirm or dispute our arithmetic.

4. The Peculiarities of Structure

As we have mentioned previously in Chapter Two, in writing the equation

$$f_i = \min_{j \neq i} [t_{ij} + f_j], \qquad i = 1, 2, \cdots, N - 1 \tag{1}$$

we are implicitly assuming either that there is a link joining every pair of vertices or, if not, that we allow only those j which are accessible from i. In Chapter Two we gave two methods for treating the problem if there are no links between some pairs of vertices. In one of these we supplied fictitious links and assigned infinite or large times to them. In the other, we introduced the concept of the connection matrix and observed that in

---

* Naturally, we numbered the cities initially in such a way that this was the case.

TABLE 3

i	Most westerly vertex accessible	Path to N	$f_i^{(1)}$
1	10		
2	16	2, 16, 22, 25, 31, 35, 37, 39, 40	2480
3	6		
4			
5			
6			
7			
8			
9			
10			
11			
12			
13			
14			
15			
16	22	16, 22, 25, 31, 35, 37, 39, 40	1970
17			
18			
19			
20			
21			
22	25	22, 25, 31, 35, 37, 39, 40	1780
23			
24			
25	31	25, 31, 35, 37, 39, 40	1590
26			
27			
28			
29	34	29, 34, 37, 39, 40	1385
30	35	30, 35, 37, 39, 40	1290
31	35	31, 35, 37, 39, 40	1230
32	34	32, 34, 37, 39, 40	1280
33	36	33, 36, 38, 40	555
34	37	34, 37, 39, 40	950
35	37	35, 37, 39, 40	810
36	38	36, 38, 40	370
37	39	37, 39, 40	630
38	40	38, 40	110
39	40	39, 40	150

place of (1) we can write the equation

$$f_i = \min_{w_{ij} \neq 0} [w_{ij} + f_j], \qquad i = 1, 2, \cdots, N - 1 \tag{2}$$

where $w_{ij} = t_{ij}$ if t_{ij} is defined, zero otherwise. This notation indicates that one allows all j such that the number w_{ij} is different from zero.

The method of fictitious links and the method based on (2) require the same amount of computer storage, since in both cases a full N by N matrix has to be stored. However, the two methods may require different amounts of computation time. The reason is that when using (2) we compute the sum $t_{ij} + f_j$ only for some values $j \neq i$, not all such j. On the other hand, it is necessary in determining which j to use to call forth all the numbers w_{ij} ($j \neq i$) from storage and test to see which have the value 0. Thus the amount of arithmetic is reduced, but the number of decision operations is increased.

Another method for keeping track of the structure of the graph is to determine the set of points accessible from each vertex i by a link. We call this set $S(i)$. If the vertices in $S(i)$ are listed in computer storage for each i, we can use the equations

$$f_i = \min_{j \in S(i)} [t_{ij} + f_j], \qquad i = 1, 2, \cdots, N - 1 \tag{3}$$

in place of (1), where the notation indicates that one allows all j in the set $S(i)$, and no other j. A computer program can be based on Eq. (3) rather than on Eq. (2); that is, on storing the sets $S(i)$ rather than the matrix (w_{ij}). This change can be expected to reduce the computation time considerably. The reason is that the decision process based on examining the w_{ij} is entirely eliminated. Of course, in return, it is necessary that the sets $S(i)$ be known and stored. In any given problem, one must select whatever method appears best in the light of both the size of the problem and the storage capacity and speed of the machine to be used. It is interesting to note that significant scheduling and optimization problems arise in the very process of using the computer to carry out an algorithmic process.

Exercises

1. Let i be fixed and consider the calculation of $[t_{ij} + f_j]$ by means of Eq. (1). Assume that fictitious links with large times have been supplied and that the numbers t_{ij} and f_j are in storage. How many additions and how many comparisons of two numbers are required to calculate the minimum?

2. For a fixed i, let k_i denote the number of nonzero numbers w_{ij}. Assuming that the numbers w_{ij} and f_j are in storage, how many additions

and comparisons must be performed by the computer in order to find the nonzero w_{ij} and to compute min $[t_{ij} + f_j]$ by the method given in (2)? If one addition takes the same time as one comparison (this will depend on the computer being used), show that method (2) is faster than the method given in (1) if $2k_i < N - 2$.

5. Computer Program

A computer program based on Eq. (4.3) could be designed as follows: We store the number, N, of vertices. For each vertex I, $I = 1, 2, \cdots, N$, we store the number of vertices accessible from I and the list of these vertices. Call the former $NO(I)$ and the latter $S(I, L)$, where $L = 1, 2, \cdots, NO(I)$. Finally, store the time $T(I, J)$ for J in the set $S(I, L)$.

For example, for the Grand Canyon problem we have the following incomplete table of data.

I	$NO(I)$	$S(I, 1)$	$T(I, 1)$	$S(I, 2)$	$T(I, 2)$	$S(I, 3)$	$\cdots$
1	3	2	220	3	255	10	$\cdots$
2	6	1	220	3	105	6	$\cdots$
3	4	1	255	2	105	4	$\cdots$
$\cdots\cdots$							
39	3	37	480	38	330	40	$\cdots$

In addition to the above information, we also store an initial approximation vector $F(J)$. Then for $I = 1, 2, \cdots, N - 1$, we calculate

$$\text{FMIN}(I) = \min_J [T(I, J) + F(J)] \qquad (1)$$

Here J must take on the values stored in $S(I, L)$ for $L = 1, 2, \cdots, NO(I)$. The rest of the program is similar to that in § 19 of Chapter Two.

Exercises

1. For the map in Fig. 3 of Chapter One, construct a complete table of the above kind.

2. Write a computer program based on Eq. (4.3) and use it to solve:
   (a) the shortest route problem for the map of Fig. 3 of Chapter One, and
   (b) the Grand Canyon problem.

6.   Storage Considerations

The iterative techniques described in the previous pages are theoretically sound, and the examples in the text show conclusively that they are feasible for maps of the size we have been considering. What happens as N begins to increase in magnitude, say when $N = 1,000$? Consider the equations

$$f_i^{(k+1)} = \min_{j \neq i} [t_{ij} + f_j^{(k)}], \qquad i = 1, 2, \cdots N - 1$$
$$f_N^{(k+1)} = 0 \tag{1}$$

and assume that each t_{ij} exists. In other to calculate the $N - 1$ values, $\{f_i^{(k+1)}\}$, $i = 1, 2, \cdots, N - 1$, using (1), we require the set of numbers $[t_{ij}]$, together with the values of the previous approximation, $\{f_i^{(k)}\}$. As i and j range independently over the values $1, 2, \cdots, N$, t_{ij} assumes N^2 possible values, supposing for the sake of generality that $t_{ij} \neq t_{ji}$. Since $t_{ii} = 0$, we see that $N^2 - N$ values must be stored. The set $\{f_i^{(k)}\}$ adds N additional quantities. Let us then think in terms of retaining N^2 values in the rapid access storage of the computer.*

If $N = 1,000$, then $N^2 = 1,000,000$, a capacity available with the largest of contemporary computers. If $N = 1,000,000$, we must look further into the future. How far is not clear.

These simple calculations show conclusively that new ideas must be introduced if we wish to maintain the claim that the algorithm we have been using remains feasible as N reaches the values cited above. As we shall see below, there are other cogent reasons for wanting to solve large-scale routing problems. These new problems of large dimension have nothing to do with puzzles, or with determining shortest routes on maps. But it is characteristic of mathematics that the same equation can describe many different types of physical phenomena, and further, that methods developed to treat the equation arising in one fashion can be immediately applied to processes completely distinct from the original. This economy of intellectual effort is one of the fascinating features of mathematics.

7.   Use of Tapes

It is to be expected that we have to pay some price for the increase in magnitude of N. It is a question of doing the best we can with our limited technological and mathematical abilities to reduce this cost. However, it is only fair to the mathematical analyst to state that problems

* Sometimes this latter is called "fast memory." However, this tern so tinged with anthropomorphism is gradually being replaced by the more prosaic but descriptive expression used above, "rapid access storage."

involving enumeration of possibilities* and choice among possibilities are genuinely difficult and that relatively little is known at the present time about systematic methods for solving them.

Returning to the recurrence relation in (6.1), we note that to evaluate $f_i^{(k+1)}$ for a particular i requires the N values t_{ij}, $j = 1, 2, \cdots, N$ and the N values $f_j^{(k)}$, $j = 1, 2, \cdots, N$. Hence, if we store the matrix (t_{ij}), $i, j, = 1, 2, \cdots, N$, on tapes (virtually unlimited as far as capacity is concerned), we need only retain in the rapid access storage at any time $2N$ values, a tremendous reduction in demand on storage.

Shifting the columns

$$c_i = \begin{pmatrix} t_{i1} \\ t_{i2} \\ \vdots \\ t_{iN} \end{pmatrix} \qquad i = 1, 2, \cdots, N$$

back and forth from tape to rapid access storage to tape again will consume much more time than previously required when everything needed was in the rapid access storage. Nonetheless, this time is small enough for this to be considered a feasible attack on the original problem, particularly when the problem arises in connection with some process of significance in the economic or engineering fields.

8. Storage of Sparse Data

There are two essentially distinct ways of representing the structure of a graph or network in a computer, and two corresponding methods of programming. In one of these the connection matrix of the graph is stored in the computer and in the other the set $S(i)$ of vertices accessible from i is stored, for each i (or, alternatively, the set of vertices from which i can be reached). We can call these the *accessibility lists*. The second method is particularly useful if only a few of the entries in the connection matrix are nonzero. Such a matrix is called *sparse*, whereas matrices with only a few zero entries are called *dense* or *full*. For example, we see from Table 2 that most of the entries in the Grand Canyon problem are zeros. (The Grand Canyon is thus sparse, not full!)

In our routing problems, the matrix (t_{ij}) is very likely to be quite sparse, in the sense that many entries will be blank, since a given vertex i is usually connected to only a few vertices j. If only the entries t_{ij} for which the link from i to j exists are stored, there is likely to be a substantial saving in storage space used. Algorithms and computer programs for large routing problems have been designed on the basis of this idea, and,

---

* Problems of this nature are usually called "combinatorial".

in particular, the program in § 5 above can be written in this fashion. We refer the interested reader to the references in the Bibliography for details of such algorithms.

9. Calculation of the t_{ij}

As we have seen above, rapid access storage of the complete set of values of the t_{ij} is impossible for large N. In a number of important situations, however, there is a way of avoiding this difficulty. The idea is a very simple one, yet extremely powerful as far as computers are concerned.

In place of storing the actual values, we store instructions for computing these values. In general, this cannot be done, but in certain cases the instructions for calculating the values are so simple that we can employ this technique for avoiding storage of data.

Consider, for example, that we have N points in a plane, P_i, specified by their Euclidean coordinates, (x_i, y_i), $i = 1, 2, \cdots, N$, and that as the distance between these points we can use the usual Euclidean distance

$$d(P_i, P_j) = [(x_i - x_j)^2 + (y_i - y_j)^2]^{\frac{1}{2}} \qquad (1)$$

In this case it is clear that in place of storing the values $\{d(P_i, P_j)\}$, it is much easier to store the $2N$ numbers, x_i, y_i, together with the algorithm for calculating $d(P_i, P_j)$. Thus we can readily calculate t_{ij} as needed, given the velocity of passage from i to j. The same situation holds for points on a sphere where the distance is calculated along a great circle.

The essential point is that by storing *algorithms* we can greatly reduce the data that must be stored in the rapid access storage. The data that is stored is that necessary for carrying out the algorithm. In a number of important processes these algorithms may themselves be of considerable complexity and thus may consume an appreciable time for execution. Frequently, we encounter a hierarchy of algorithms.

In these cases we are trading time (which we possess to some reasonable extent) for rapid access storage capacity (which we possess in very limited quantity). In each case, it is a matter of economics, of evaluating the cost in time versus the value obtained from the determination of the optimal route.

10. Transformation of Systems

In the administration of the systems that constitute our world a frequent problem is that of deciding how to take a system from some initial state to some desired terminal state in an "efficient" fashion. Here efficiency is measured in terms of time, money, manpower or, generally, in terms of some combination of these resources.

Suppose for example, that a spacecraft is to be launched from a position on Earth with some initial velocity to land on the Moon with, hopefully, a zero terminal velocity, a "soft landing." As the spacecraft travels through space, the state of the system (in this case the spacecraft) at each time may be conveniently described by the three position coordinates and the three velocity coordinates. This is under the simplifying assumption

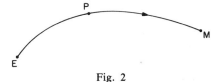

Fig. 2

that the spacecraft can be considered to be a point. If we want a more realistic description (generally necessary if we want to exercise meaningful control), we must introduce various possibilities of rotation around the axes of the spaceship, which means that the state will require many additional numbers for its description.

In any case, the motion of the object through space may be regarded as a path through "state space" (usually called "phase space" in physics).

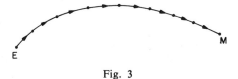

Fig. 3

The path that is actually travelled depends upon a set of decisions concerning the burning rate of fuel, use of rockets to change direction, and so forth. The choice of an optimal trajectory, one of minimum time, or minimum fuel, or maximum probability of arrival, may thus be viewed as the choice of an optimal route in phase space. The continuous path in time may be replaced by a discrete path in state space with an accuracy sufficient for all practical purposes. Once this has been done, we can use the foregoing methods to obtain the most efficient paths in many cases. In other cases we encounter serious dimensionality difficulties. Fortunately, we possess a number of different types of methods which can be successfully applied to problems of this nature which escape the foregoing methods.

Another important set of examples illustrating the significance of choosing paths through networks is furnished by economics. Consider the task of planning to increase steel production. The state of the system is now specified by a set of numbers describing, for example, the current stockpiles of steel of different types, the existing demands for this steel, and the productive capacities of steel factories of varied kinds. In order to plan effectively, it is necessary to decide how to allocate available steel to meeting the demands of consumers, to meeting the current requirements for steel production itself, to building additional steel factories, and so forth.

Analogous problems connected with the allocation of resources arise in agricultural planning, in the distribution of water resources, and generally, in the administration of the major systems of our society.

Regard the numbers used to describe the state of the system at any time as constituting the components of a point in an M-dimensional space. We see that a decision corresponds to a change in these numbers, or as we shall say, a transformation from one point in phase space to another. Associated with each decision, or transformation, is a cost, if negative, or a gain, if positive.

The problem of optimal planning may thus be regarded as that of a routing problem in a higher dimensional space. The higher the dimension of this space, which is to say the more complex the system, the more numbers are required to describe it. This, in turn, produces a greater number of vertices. This is one of the reasons that we are interested in routing problems with very large numbers of vertices.

Processes of this type are often of the kind described in the foregoing section where the cost of going from point i to point j can be calculated when needed without the necessity for storing it. This fact often enables us to bypass one part of the dimensionality roadblock.

11. Stratification

In a number of planning processes the geometric configuration is such that we can avoid the use of successive approximations completely and determine the f_i in a direct fashion. Simultaneously, we can greatly reduce the demands on rapid access storage. Consider, for example, a map of the following type:

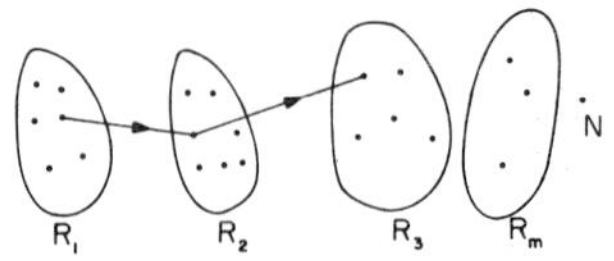

Fig. 4

The requirement is that we start at a vertex in region one, R_1, and go to some vertex in R_2, from this vertex in R_2 to some vertex in R_3, and so on, until we get to N, the terminal point.

Using the equation

$$f_i = \min_{j \neq i} (t_{ij} + f_j) \qquad (1)$$

it is clear that f_i is determined immediately for the points $i = k_{M-1} + 1$, $\cdots, k_M$ in R_M since we must go directly from any point in R_M to N. Once these values have been determined, we use (1) to obtain the values of f_i for points $i = k_{M-2} + 1, \cdots, k_{M-1}$ in R_{M-1}, and so on, until we work our way back to R_1.

Sometimes this stratification is apparent from the map or the underlying physical process. For example, R_1 may be the set of possible states at time t_1, R_2 the states at time t_2, etc., where $t_1 < t_2 \cdots$. In other cases, some ingenuity is required to obtain a decomposition of the foregoing type.

12. Acceleration of Convergence

Thus far we have employed the recurrence relation

$$f_i^{(k)} = \min_{j \neq i} [t_{ij} + f_j^{(k-1)}], \qquad i = 1, 2, \cdots, N - 1$$
$$f_N^{(k)} = 0 \qquad (1)$$

with $f_i^{(0)}$ an initial guess. That is, from the values $f_1^{(k-1)}, \cdots, f_N^{(k-1)}$, we determine $f_1^{(k)}$, then $f_2^{(k)}$, and so on.

Presumably $f_1^{(k)}$ is a better estimate for f_1 than $f_1^{(k-1)}$, $f_2^{(k)}$ is a better estimate for f_2 than $f_2^{(k-1)}$, and so on. With this in mind we can accelerate the convergence of the foregoing approximation process by using the best current estimates for $f_1, f_2, \cdots, f_{N-1}$ to calculate $f_i^{(k)}$. Thus, in place of (1) we write

$$f_i^{(k)} = \min [t_{i1} + f_1^{(k)}, \cdots, t_{i, i-1} + f_{i-1}^{(k)},$$
$$t_{i, i+1} + f_{i+1}^{(k-1)}, \cdots, t_{i, N-1} + f_{N-1}^{(k-1)}, t_{iN}] \qquad (2)$$

For example consider again Fig. 12 of Chapter Two repeated below as Fig. 5, with times t_{ij} as shown. The results of applying Eq. (1) and

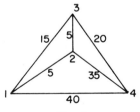

Fig. 5

Eq. (2) are shown in Table 4. The results using best current estimates (sometimes called "updating") and computing in the order 4, 3, 2, 1 are obtained successively as follows:

$$f_3^{(2)} = \min\ (t_{31} + f_1^{(1)},\ t_{32} + f_2^{(1)},\ t_{34})$$
$$= \min\ (15 + \infty,\ 5 + \infty,\ 20) = 20$$

$$f_2^{(2)} = \min\ (t_{21} + f_1^{(1)},\ t_{23} + f_3^{(2)},\ t_{24})$$
$$= \min\ (5 + \infty,\ 5 + 20,\ 35) = 25 \qquad (3)$$

$$f_1^{(2)} = \min\ (t_{12} + f_2^{(2)},\ t_{13} + f_3^{(2)},\ t_{14})$$
$$= \min\ (5 + 25,\ 15 + 20,\ 40) = 30$$

Notice that in computing $f_2^{(2)}$ we use $f_3^{(2)}$, the just-computed value, rather than $f_3^{(1)}$, and in computing $f_1^{(2)}$ we use both $f_2^{(2)}$ and $f_3^{(2)}$. In this way, we obtain convergence in one step rather than three steps as required if we use Eq. (1).

On the other hand, if we use best current estimates but compute in the order 1, 2, 3, 4, we gain no advantage. It is generally desirable to use

TABLE 4

ITERATION USING (1)

i	$f_i^{(1)}$	i	$f_i^{(2)}$	i	$f_i^{(3)}$	i	$f_i^{(4)}$
1	∞	4	40	3	35	2	30
2	∞	4	35	3	25	3	25
3	∞	4	20	4	20	4	20
4	0	—	0	—	0	—	0

ITERATION USING UPDATING (IN ORDER 1, 2, 3, 4)

i	$f_i^{(1)}$	i	$f_i^{(2)}$	i	$f_i^{(3)}$	i	$f_i^{(4)}$
1	∞	4	40	3	35	2	30
2	∞	4	35	3	25	3	25
3	∞	4	20	4	20	4	20
4	0	—	0	—	0	—	0

ITERATION USING UPDATING (IN ORDER 4, 3, 2, 1)

i	$f_i^{(1)}$	i	$f_i^{(2)}$
1	∞	2	30
2	∞	3	25
3	∞	4	20
4	0	—	0

best current estimates and to work in the order of increasing distance or time from the destination.

As we saw in § 23 of Chapter Two, the labelling algorithm provides a convenient way of carrying through the iteration in Eq. (1) by hand. The accelerated scheme in Eq. (2) likewise corresponds to an accelerated labelling algorithm, which we can describe as follows.

Step 1. Choose an order in which to consider the vertices (for example, N, $N-1, \cdots, 2, 1$).

Step 2. Write a zero beside the terminal vertex, N, and ∞ beside the others.

Step 3. For each vertex i in order $(i \neq N)$ compute the sum of t_{ij} and the label on vertex j, for each $j \neq i$. Compute the minimum m_i of these sums and record this as the label on vertex i.

Step 4. Return to step 3; repeat step 3 until the labels no longer change.

Methods similar to (2) have long been used in iteration schemes for solving linear algebraic equations. Sometimes these are called "Seidel methods."

The acceleration or updating technique can be applied in conjunction with any of the methods previously described for reducing storage requirements or the amount of calculation.

Exercises

1. Use the accelerated convergence method (2) for the map in Fig. 15 of Chapter Two. Take the vertices in the order
   (a) 1, 2, 3, 4, 5, 6, 7, 8; and
   (b) 8, 5, 4, 6, 7, 3, 2, 1.

2. Write a computer program based on Eq. (2). Assume that (t_{ij}) is a full N by N matrix to be placed in rapid access storage.

3. Write a program which is based on the equations
$$f_i = \min_{j \in S(i)} [t_{ij} + f_j]$$
and which uses the most current value of f_j (see § 5). .Assume that the calculation is to proceed in the order $i = N, N-1, \cdots, 2, 1$.

4. Explain how to use the acceleration idea in conjunction with the method of Chapter Two, § 24, and illustrate for the map discussed there.

5. Write a program for the method of Chapter Two, § 24, Eq. (5).

6. Solve the Grand Canyon problem in each of these ways and compare for efficiency:

(a)   Using the program of Exercise 3 with initial policy

$$f_i^{(1)} = \infty, \qquad f_N^{(1)} = 0$$

(b)   Using the program of Exercise 3 with initial policy in Table 3.

(c)   Using the program of Exercise 5.

(d)   Using the program for Dijkstra's algorithm (miscellaneous Exercise 3 in Chapter Two).

13.   Design of Electric Power Stations

Electric generating stations contain large systems of cables for the distribution of power within the station. These cables are run through "cable trays," which are tubes, ordinarily of rectangular cross-section, which provide a common housing for a number of cables. The tray system is composed of horizontal runs, usually hung just beneath each floor, and vertical runs between floors. In Fig. 6, we illustrate a possible tray layout. The building outlines are shown by straight lines and the cable trays by dotted areas. Several terminals are indicated by the letters A, B, C, etc.

It is clear that a cable from A to E, for example, can be laid through the trays along several different routes. One of the problems facing the engineer is to select a route for each cable, given the list of all connections which are to be made and given the locations of all trays and the lengths of these trays between branch points. In order to minimize cost, it is desirable to select a path for each cable which minimizes the length of cable used.

By assigning a vertex number to each terminal and to each point where trays intersect, we can replace the physical diagram by a graph of the type we have been discussing. We have done this for Fig. 6, the result being Fig. 7. The lengths assigned to the various edges in the latter graph are the physical lengths of the cable trays from Fig. 6, and are not shown to scale on the graph. These could be entered in a matrix t_{ij} just as before. The problem of determining the shortest route for a cable from terminal A to terminal E now becomes the problem of determining a shortest path from vertex 1 to vertex 10 on the graph.

In practice, this problem becomes very complicated since it is necessary to take into account different sizes of cable and of cable trays, and various other factors. Nevertheless, a successful computer program has been designed and used for this purpose. For more details, see the reference given in the Bibliography and Comments.

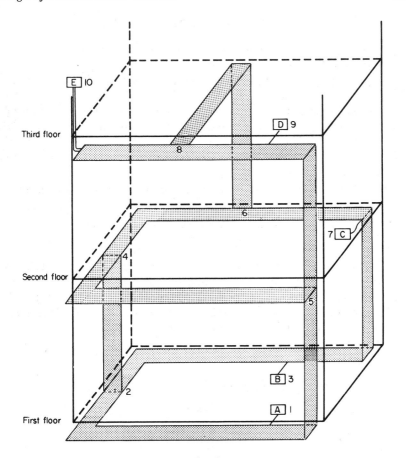

Fig. 6

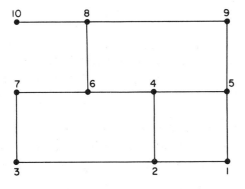

Fig. 7

Miscellaneous Exercises

1. Suppose that the times to traverse a given section of road are not constant, but depend on the particular time at which one reaches the beginning of the section. For example, suppose the time required depends on the time of day, a realistic supposition in view of the fluctuation in the amount of traffic. Formulate the routing problems for this case, introducing appropriate minimal time functions $f_i(t)$ which depend on the time t of starting from i. Find the equations corresponding to (1.1) and discuss how the method of successive approximations might be applied here.

   One formulation and solution of this problem has been given by

   K. L. Cooke and E. Halsey, "The Shortest Route Through a Network with Time-Dependent Internodal Transit Time," *J. Math. Anal. Appl.* **14** (1966) 493–498.

   Compare your solution with theirs.

2. The shortest route problem can also be formulated as a "linear programming" problem. To do this, one defines variables x_{ij} ($i, j = 1, 2, \cdots, N$) which take only the values 0 and 1. For each possible path P from the starting vertex 1 to the terminal vertex N these variables take on well-defined values, as follows:

$$x_{ii} = 1 \text{ if the path goes through vertex}$$
$$i \ (i = 1, \cdots, N)$$
$$= 0 \text{ otherwise}$$
$$x_{ij} = 1 \text{ if the path uses the link from } i \text{ to } j$$
$$(i \neq j; \ i, j = 1, \cdots N)$$
$$= 0 \text{ if it does not}$$

   The optimal path is then the one for which the sum

$$S = \sum_{i=1}^{N} \sum_{j=1}^{N} t_{ij} \, x_{ij}$$

   is least. The x_{ij} must satisfy certain constraints to represent the fact that a path which enters a vertex must leave it. This is a typical linear programming problem. Write out the complete set of equations and the sum S for one of the examples of Chapter One or Two. For further details see

   G. B. Dantzig, *Linear Programming and Extensions*, Princeton University Press.

   G. B. Dantzig, "Discrete-Variable Extremum Problems," *Oper. Research* **5** (1957) 266–277.

3. Extend Dijkstra's algorithm to the problem treated in Exercise 1. See

   S. Dreyfus, *An Appraisal of Some Shortest Path Algorithms*, RAND Corporation, RM-5433-PR, Oct. 1967, Santa Monica, Calif.

4. Consider the routing problem under the additional condition that the path must pass through k specified nodes. See

J. P. Saksena and S. Kumar, "The Routing Problem with k Specified Nodes," *Oper. Research* **14** (1966) 909–913.

S. Dreyfus, *op. cit.*

R. Kalaba, "On Some Communication Network Problems," *Combinatorial Analysis, Proc. Symp. Appl. Math.* **10** (1960) 261–280.

Bibliography and Comments

The algorithms which we have emphasized, both the labelling methods and the methods based on recurrence equations, are sometimes called "tree-building" algorithms, since the set of shortest paths to a fixed destination is a tree. Other methods, called "matrix methods," find shortest paths between all pairs of vertices simultaneously. These methods can be very fast, but since our primary aim has been to introduce ideas associated with computation rather than to give a comprehensive survey, we have elected not to discuss these methods. The interested reader may consult the survey paper of Dreyfus previously cited and the following papers:

R. W. Floyd, "Algorithm 97, Shortest Path," *Comm. A.C.M.* **5** (1962) 345.

G. B. Dantzig, *All Shortest Routes in a Graph*, Operations Research House, Stanford University, Technical Report No. 66–3, 1966.

B. A. Farbey, A. H. Land, and J. D. Murchland, "The Cascade Algorithm for Finding all Shortest Distances in a Directed Graph," *Management Science* **14** (1967) 19–28.

One of the difficulties associated with the set of matrix methods is that the matrix of the graph must be stored and operations performed on it. As we have seen, this may be impossible with contemporary computers if the number of vertices is large. If the matrix is sparse, however, it may be possible to "tear" the graph into several subgraphs in such a way that there are few connections between different subgraphs. This decomposition, similar in spirit to the stratification discussed above, can produce a great saving in computation time. See

G. Mills, "A Decomposition Algorithm for the Shortest-Route Problem," *Oper. Research* **14** (1966) 279–291.

A. H. Land and S. W. Stairs, "The Extension of the Cascade Algorithm to Large Graphs," *Management Science* **14** (1967) 29–33.

T. C. Hu, "A Decomposition Algorithm for Shortest Paths in a Network," *Oper. Research* **16** (1968) 91–102.

T. C. Hu and W. T. Torres, *A Short Cut in the Decomposition Algorithm for Shortest Paths in a Network*, Mathematics Research Center, U.S. Army, University of Wisconsin, MRC Technical Summary Report No. 882, July, 1968.

Efficient computer algorithms can be found in:

A. Perko, "Some Computational Notes on the Shortest Route Problem," *Computer J.* **8** (1965) 19–20.

U. Pape, "Some Computational Notes on the Shortest Route Problem," *Computer J.* **11** (1968) 240.

T. A. J. Nicholson, "Finding the Shortest Route Between Two Points in a Network," *Computer J.* **9** (1966) 275–280.

G. Mills, "A Heuristic Approach to Some Shortest Route Problems," *CORS J.* **6** (1968) 20–25.

The following paper treats the problem of accelerating convergence, particularly for so-called *reticular networks*:

F. Luccio, "On Some Iterative Methods for the Determination of Optimal Paths Through a Network," *Calcolo* **3** (1966) 31–48.

Let us also add some further comments. The production processes discussed briefly in § 10 are sometimes called "bottleneck processes." An introduction to their study may be found in

R. Bellman, *Dynamic Programming*, Princeton University Press, 1957.

The stratification techniques discussed in § 11 can be used to treat checkers and endgames in chess. See

R. Bellman, "On the Application of Dynamic Programming to the Determination of Optimal Play in Chess and Checkers," *Proc. Nat. Acad. Sci.* **53** (1965) 244–247.

R. Bellman, "Stratification and Control of Large Systems with Applications to Chess and Checkers," *Information Sciences* **1** (1968) 7–21.

The systematic use of "tearing" method was introduced by G. Kron in connection with electric networks. See

J. P. Roth, "An Application of Algebraic Topology, Kron's Method of Tearing," *Q. Appl. Math.* **17** (1959) 1–23.

For a further discussion of the results in § 14, see

L. E. Ruhlen and P. R. Shire, *Optimal Circuit Routing by Dynamic Programming*, IEEE 5th Power Industry Computer Applications Conference, May, 1967.

Chapter   Four

PROOF OF THE VALIDITY OF
THE METHOD

1.   Will the Method Always Work?

In the foregoing pages we have applied the method of successive approximations in various ways to find a numerical solution of the system of equations

$$f_i = \min_{j \neq i} [t_{ij} + f_j], \qquad i = 1, 2, \cdots, N-1$$
$$f_N = 0 \tag{1}$$

for a number of specific maps.   In our desire to present the method in simple terms and to show how to apply it, we have avoided two questions of fundamental importance:

(a)   How do we know that the method always yields a solution of (1)?

(b)   How do we know that a solution of this equation enables us to solve the original problem of determining a shortest route?

We have previously indicated by means of examples in Chapter Two that the method of successive approximations may fail when applied to the solution of simple algebraic equations.   Consequently, the reader is already aware of the importance of establishing the validity of any particular algorithm.   The point of the second question is that the equations in (1) may have many solutions, and it is conceivable that our method produces an extraneous solution; i.e., one which has no relation to the original shortest route problem.   Hence, it is essential to establish uniqueness of solution.   Furthermore, we wish to discuss precisely how the solution of the equation

123

is used to solve the original problem.

The reader is familiar with the concept of extraneous solution from his courses in algebra. Let us recall a way in which such a solution occurs. Suppose that we are informed of the existence of a rectangular lot with the property that the length exceeds the width by ten feet, and that the area is six hundred square feet. We are then asked to determine the dimen-

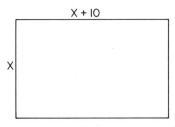

X + 10

X

Fig. 1

sions. Proceeding in a familiar fashion, we let x denote the width. Then the conditions of the problem yield the quadratic equation

$$x(x + 10) = 600 \qquad (2)$$

The numbers have been carefully chosen so that we have a simple factorization

$$x(x + 10) - 600 = (x - 20)(x + 30) \qquad (3)$$

Hence, we see that $x = 20$, $x + 10 = 30$, and thus that the lot is 20×30.

What about the other solution $x = -30$? We smile and say that this is obviously a nonphysical solution. It has no meaning as far as the original problem is concerned. This is, however, a disturbing example. Even though, as in the foregoing case, the verbal problem possesses exactly one solution, the corresponding analytic problem may possess more than one solution.

Here, the negative value immediately alerts us to the fact that the second solution is spurious. How can we be certain, however, that there are not positive solutions of the equation of interest to us, (1), which are equally spurious? Questions of this type are of the utmost importance in all applications of mathematics and particularly so in connection with the use of a digital computer. It is often not difficult to write a computer program that will use tens of hours to produce millions of numbers. It is entirely another matter to ensure that these numbers are meaningful.

When confronted by a formidable equation, there are two questions that demand priority:

(a)   Does the equation have at least one solution?

(b)   Does the equation possess more than one solution?

The first is a matter of *existence*; the second a matter of *uniqueness*.

It is clear that there is little point to trying to obtain a numerical solution if the equation does not possess a solution. On the other hand, if there are several solutions, it is vital to ascertain which solution is being calculated, and which solution corresponds to the answer to the original physical problem. In the following pages, we will discuss these matters in some detail. To our relief, it will turn out that the method presented in the earlier chapters always yields the desired answer to the original problem of optimal routing. Furthermore, the proof of this assertion is, equally to our relief, quite simple.

2. Is There at Least One Solution?

In this section, we want to show that (1.1) possesses at least one solution—the solution corresponding to the answer to the original routing problem.

Consider the question of determining the shortest route from the ith vertex to the terminal point N. There are only a finite number of ways of going from i to N without passing through the same vertex twice. Since we are trying to minimize, it is clear that we are only interested in paths of this type. The problem of determining the shortest route is therefore of finding the smallest of a finite set of positive numbers, where each number is the length of a permissible path. This minimum value certainly exists, and clearly depends on the point i at which we start. Let g_i, $i = 1, 2, \cdots, N - 1$, denote the minimum value, and set $g_N = 0$.

We wish to show that the g_i satisfy (1.1). To this end, we employ a familiar argument, essentially a repetition of previous arguments. In the minimal path starting from i, we must go to some other point, say j. Obviously, the continuation from j must be minimal. The proof of this last statement is by contradiction. Hence, for some j,

$$g_i = t_{ij} + g_j \tag{1}$$

Since g_i is itself the minimum time required to go from i to N, we see that j must be chosen to minimize the expression on the right in (1). The proof of this statement is, as before, by contradiction. Hence,

$$g_i = \min_{j \neq i} [t_{ij} + g_j], \qquad i = 1, 2, \cdots N - 1 \tag{2}$$

the desired result.

3. How Do We Find the Desired Solution?

Now that we have established that the desired answer is included among the solutions of (1.1), let us next show that the various methods of successive approximations we have employed produce this answer.

Let us recall one method we employed. We introduced first the quantities

$$f_i^{(0)} = t_{iN}, \qquad i = 1, 2, \cdots, N-1 \tag{1}$$

with $f_N^{(0)} = 0$. Then we considered the quantities $\{f_j^{(k)}\}$, defined by the relations

$$f_i^{(k)} = \min_{j \neq i} [t_{ij} + f_j^{(k-1)}] \tag{2}$$

$f_N^{(k)} = 0$, for $k = 1, 2, \cdots$.

Let us recall the meaning of the quantities calculated in this fashion. The quantities $f_i^{(0)}$ are the times required when we follow the naive policy of going directly from i to N. The $f_i^{(1)}$ are the minimum times required when we allow at most one intermediate stop in going from i to N. Generally, the $f_i^{(k)}$ are the minimum times required when at most k intermediate stops are allowed (see Chapter Two, § 22).

Since a path of minimum time will require at most $N-2$ intermediate stops (since no return to a vertex previously visited is desirable), it is clear that the numbers $f_i^{(k)}$ will eventually start repeating themselves, that is, from some value of k on we shall have $f_i^{(k)} = f_i^{(k+1)} = f_i^{(k+2)} = \cdots$. As a matter of fact, we see that this repetition must take place before $k = N - 1$.

Thus the method of successive approximations employed here converges in a *finite number of steps* to the desired solution. It is this fact which enables us to bypass the classical theory of convergence of infinite sequences.

4. Monotonicity

Although we have now proved convergence to the desired values of the successive iterates when the initial approximation is given by (3.1), we have not yet proved that other initial approximations (such as those used in some of the examples in the previous chapters) yield convergent sequences of iterates. In the next few sections, this will be proved with the aid of certain "monotonicity" results.

It is clear from the physical meaning of the $f_i^{(k)}$ in (3.2) that

$$f_i^{(0)} \geq f_i^{(1)} \geq \cdots \geq f_i^{(k)} \geq f_i \tag{1}$$

where f_i, $i = 1, 2, \cdots, N-1$, denotes the solution to the original routing problem. These relations can be proved analytically, without reference to the physical meaning of the quantities $f_i^{(k)}$, as follows. Comparing (3.1) and (3.2), we see that

$$f_i^{(1)} = \min_{j \neq i} [t_{ij} + f_j^{(0)}] \leq t_{iN} = f_i^{(0)}, \qquad i = 1, 2, \cdots, N - 1 \qquad (2)$$

since $j = N$ is an admissible choice, and $f_N^{(0)} = 0$. Since this inequality holds, we have

$$f_i^{(2)} = \min_{j \neq i} [t_{ij} + f_j^{(1)}] \leq \min_{j \neq i} [t_{ij} + f_j^{(0)}] = f_i^{(1)} \qquad (3)$$

We now proceed inductively to show that $f_i^{(k)} \leq f_i^{(k-1)}$ for $k \geq 1$ and for $i = 1, 2, \cdots, N - 1$.

To show that $f_i^{(k)} \geq f_i$ for each k, we proceed inductively once again. We have

$$f_i = \min_{j \neq i} [t_{ij} + f_j] \leq t_{iN} \qquad (4)$$

using the fact that $j = N$ is an admissible choice and that $f_N = 0$. We see then that $f_i \leq f_i^{(0)}$. Hence, using (4) again,

$$f_i = \min_{j \neq i} [t_{ij} + f_j] \leq \min_{j \neq i} [t_{ij} + f_j^{(0)}] = f_i^{(1)} \qquad (5)$$

We now can use induction to show that $f_i \leq f_i^{(k)}$ for all $i \geq 0$.

The foregoing results are, as mentioned above, immediate from the definitions of the quantities involved. However, as we have previously pointed out, it is quite important for the student to gain facility in proving "obvious" statements. Unless an obvious statement has an equally obvious proof, there is a serious lack of understanding of the underlying mathematics.

5. The Upper Solution

At the moment, we have not yet established the uniqueness of the solution we have obtained. That is, the possibility remains that (1.1) can be satisfied by a set of numbers other than those denoting the minimal times we seek. We can, however, easily establish the following intersting fact: The solution obtained following the procedure outlined in §3 dominates any other solution. By this, we mean that if g_i denotes any other solution of (1.1), then

$$f_i \geq g_i, \qquad i = 1, 2, \cdots, N - 1 \qquad (1)$$

To establish this, we use the same argument as before. From the relations

$$\begin{aligned} g_i &= \min [t_{ij} + g_j] \leq t_{iN}, \qquad i \neq N \\ g_N &= 0 \end{aligned} \qquad (2)$$

we see that $g_i \leq f_i^{(0)}$. Hence,

$$g_i = \min_{j \neq i} [t_{ij} + g_j] \leq \min_{j \neq i} [t_{ij} + f_j^{(0)}] = f_i^{(1)} \qquad (3)$$

Thus, inductively, we see that

$$g_i \leq f_i^{(k)} \tag{4}$$

for $k = 0, 1, 2, \cdots$ Since $f_i^{(N-1)} = f_i$, we have the desired inequality.

6. The Lower Solution

Let us use similar ideas to obtain a solution of (1.1) which is, in turn, dominated by any other solution. Let us consider the method of successive approximations initiated by setting

$$F_i^{(0)} = \min_{j \neq i} t_{ij}, \qquad i = 1, 2, \cdots N - 1, \qquad F_N^{(0)} = 0 \tag{1}$$

This corresponds to going to the nearest point from i. Then set

$$F_i^{(1)} = \min_{j \neq i} [t_{ij} + F_j^{(0)}], \qquad i = 1, 2, \cdots, N - 1, \qquad F_N^{(1)} = 0 \tag{2}$$

and, generally, set

$$F_i^{(k+1)} = \min_{j \neq i} [t_{ij} + F_j^{(k)}], \qquad i = 1, 2, \cdots, N - 1, \\ F_N^{(k+1)} = 0 \tag{3}$$

We assert that

$$F_i^{(0)} \leq F_i^{(1)} \leq \cdots \leq F_i^{(k)} \leq \cdots \tag{4}$$

This is established inductively as in the previous sections.

Further, let us show that if g_i denotes any solution of (1.1) for which $g_1, \cdots, g_N$ are nonnegative then

$$F_i^{(k)} \leq g_i \tag{5}$$

for each k. The result is true for $k = 0$ since

$$F_i^{(0)} = \min_{j \neq i} t_{ij} \leq \min_{j \neq i} [t_{ij} + g_j] = g_i \tag{6}$$

As above, the result now follows inductively.

Next we wish to establish the fact that the sequence $\{F_i^{(k)}\}$ converges in a finite number of steps. Our first impulse is to assert that it must converge since it corresponds to actual paths among the vertices. For example, $F_i^{(1)} = t_{ij} + F_j^{(0)} = t_{ij} + t_{jk}$ for some choice of j and k, so that $f_i^{(1)}$ is the time for a certain two-link path (see § 11 below). Since there are only a finite number of possible paths, any monotone increasing sequence of times should eventually repeat itself, and thus converge in a finite number of steps.

Unfortunately, considering the way in which the sequence $\{F_i^{(k)}\}$ has been generated there is nothing to guarantee that no loops occur in the paths associated with $F_i^{(k)}$. If there is no restriction on the number of such loops, there are infinitely many possible paths. However, since the $F_i^{(k)}$ are uniformly bounded by the f_i, as evidenced by (5), and since every $t_{ij} > 0$, it is evident that the number of loops is uniformly bounded. By this we mean that there is a number b such that no path from any vertex can loop more than b times.

With this additional information, we can apply the foregoing argument. There are only a finite number of possible paths with a uniformly bounded number of loops. Consequently, any monotone increasing sequence of times for these paths must eventually settle down to the same value. Let F_i denote this value, $i = 1, 2, \cdots N - 1$.

Having established convergence, we now deduce from (5) that

$$F_i \leq g_i, \qquad i = 1, 2, \cdots, N - 1 \tag{7}$$

for any other solution $g_i \geq 0$. Thus, any nonnegative solution g_i is constrained to lie between F_i and f_i, $F_i \leq g_i \leq f_i$.

7. Experimental Proof of Uniqueness

We now possess a very simple way of establishing the uniqueness of the solution of (1.1) in any particular case. We calculate the two functions F_i and f_i, using the algorithms described above. If it turns out that $F_i = f_i$, $i = 1, 2, \cdots, N - 1$, then we have demonstrated uniqueness, by virtue of the relation $F_i \leq g_i \leq f_i$.

This is an important idea in connection with the development of contemporary computers. In many situations we can develop ad hoc procedures of the foregoing type which can be tested experimentally. In particular, there are many cases where the success of the method of successive approximations depends upon the choice of the initial approximation and where only the calculation itself can tell us whether the choice was judicious or not. The concept of upper and lower solutions plays an important role in other areas of analysis.

8. Uniqueness of Solution

We shall now prove analytically that there cannot be two distinct solutions of (1.1). In order to do this, we shall assume that g_i $(i = 1, 2, \cdots, N)$ and h_i $(i = 1, 2, \cdots, N)$ are two solutions and we shall prove that $g_i = h_i$ $(i = 1, 2, \cdots, N)$, thus demonstrating that any two solutions turn out to be identical.

We know that $g_N = 0$ and $h_N = 0$. If $g_i = h_i$ for every i, there is nothing to be proved. Hence, we start with the hypothesis there is at least one value of i for which g_i and h_i are different. Looking at all values of i for which $|g_i - h_i|$ is different from zero (if any), we now pick out the index i for which this difference is largest. By changing the numbering of the vertices, if necessary, we can suppose that $i = 1$ gives the largest difference; and by interchanging the names of g_i and h_i for every i, if necessary, we can suppose that $g_1 - h_1 > 0$. We now have

$$g_1 - h_1 \geqq g_i - h_i, \qquad i = 2, 3, \cdots, N \tag{1}$$

On the other hand, from (1.1) we see that

$$\begin{aligned} g_1 &= \min_{j \neq i} (t_{1j} + g_j) \\ h_1 &= \min_{j \neq i} (t_{1j} + h_j) \end{aligned} \tag{2}$$

Let us suppose that a value of j giving the minimum in the second equation is 2, which we can arrange by renumbering the vertices 2 to N, if necessary. Then we have

$$\begin{aligned} g_1 &\leqq t_{1j} + g_j, \quad \text{for} \quad j = 2, 3, \cdots, N \\ h_1 &= t_{12} + h_2, \quad h_1 \leqq t_{1j} + h_j, \quad \text{for} \quad j = 3, 4 \cdots, N \end{aligned} \tag{3}$$

These relations lead us to

$$g_1 - h_1 \leqq (t_{12} + g_2) - h_1 = g_2 - h_2 \tag{4}$$

Combining this inequality with (1), we see that $g_1 - h_1 = g_2 - h_2$.

Now we repeat this argument. We have

$$\begin{aligned} g_2 &= \min_{j \neq 2} (t_{2j} + g_j) \\ h_2 &= \min_{j \neq 2} (t_{2j} + h_j) \end{aligned} \tag{5}$$

The value of j giving the minimum in the second equation cannot be $j = 1$, since

$$t_{21} + h_1 = t_{21} + t_{12} + h_2 > h_2 \tag{6}$$

We can therefore suppose that it is $j = 3$ (renumbering the vertices 3 to N if necessary). Then we have

$$\begin{aligned} g_2 &\leqq t_{2j} + g_j, \quad j = 1, 2, \cdots, N \\ h_2 &= t_{23} + h_3, \quad h_2 \leqq t_{2j} + h_j, \quad \text{for} \quad j = 1, 2, \cdots, N \end{aligned} \tag{7}$$

Hence

$$g_1 - h_1 = g_2 - h_2 \leqq t_{23} + g_3 - h_2 = g_3 - h_3 \tag{8}$$

Using (1), we therfore see that $g_1 - h_1 = g_3 - h_3$. Thus, $g_1 - h_1 = g_2 - h_2 = g_3 - h_3$.

By continuing in this manner for $N-1$ steps, we arrive at the continued equation $g_1 - h_1 = g_2 - h_2 = \cdots = g_i - h_i$ $(i = 1, 2, \cdots, N-1)$, thus proving that the two solutions are in fact identical.

We have previously shown that the desired minimal times comprise a solution of (1.1). Thus it follows from the uniqneness just proved that if we can by any method whatsoever find a solution $f_1, \cdots, f_{N-1}$ of (1.1) then this solution provides us with the desired minimal times.

9. Determination of Optimal Policy

The original problem is that of going from an initial point, 1, to a terminal point, N, in shortest time. A solution is a set of numbers which determine the path which yields this minimal time. In order to solve this particular problem, we imbedded it within the general problem of obtaining the path of minimum time from an arbitrary point, i, $i = 1, 2, \cdots$, $N-1$.

Once we take this point of view, we can present the solution in a simpler form. In place of giving the complete path of minimum time from i to N, it is sufficient to give the rule which determines the point j that one goes to next from i. This rule is a function, $j(i)$, which we have called a *policy*. A policy which yields the minimum time is called an *optimal policy*.

Observe that the knowledge of an optimal policy determines the minimal path. Starting at i, we calculate

$$i_1 = j(i)$$
$$i_2 = j(i_1) \tag{1}$$

and so forth until we reach N. The path is then $[i, i_1, i_2 \cdots, N]$.

Hence, we can assert that an optimal policy determines the minimal times, $\{f_i\}$. Conversely, we can assert that a knowledge of the f_i, plus the information that the solution of (1.1) is unique, determines the optimal policy $j(i)$. To see this, we examine the equation

$$f_i = \min_{j \neq i} [t_{ij} + f_j] \tag{2}$$

Then $j(i)$ is obtained as the value, or values, which furnishes the minimum on the right. Note that $j(i)$ need not be single valued. In other words, there may be many paths which require the same minimum time.

The equivalence of a solution in terms of either f_i or $j(i)$ is of great significance in more advanced studies of optimization processes. It is a duality property of both analytic and computational import.

10. Arbitrary Initial Approximation

It follows from the uniqueness just established that the upper and lower solutions discussed in § 5 and 6 are identical. Moreover, we can now prove that for any initial approximation $g_i^{(0)}$ for which $g_i^{(0)} \geq 0$ ($i = 1, \cdots,$ $N - 1$), $g_N^{(0)} = 0$, the successive approximations defined by the equations

$$g_i^{(k)} = \min_{j \neq i} (t_{ij} + g_j^{(k-1)}), \qquad i = 1, 2, \cdots, N - 1$$
$$g_N^{(k)} = 0 \tag{1}$$

with $k = 1, 2, \cdots$, converge to the (unique) solution. Furthermore, what is remarkable is that the convergence occurs in a finite number of steps. The practical importance of this observation is that we can use any initial policy which is suggested to us on intuitive grounds, and be confident that we shall obtain the correct solution starting with the times furnished by this policy as an initial approximation. For example, in the original routing problem of Chapter One, we used initial policies such as "Go right if possible, otherwise up $\cdots$." Such initial policies, if adroitly selected, may yield approximations, in between those for the upper and lower solutions, which converge very rapidly.

The proof of convergence of $g_i^{(k)}$ is as follows. Let $f_i^{(k)}$ be defined as in § 3 (the upper solution) and let $F_i^{(k)}$ be defined as in § 6 (the lower solution). We have

$$g_i^{(1)} = \min_{j \neq i} (t_{ij} + g_j^{(0)}) \leq t_{iN} + g_N^{(0)} = t_{iN} = f_i^{(0)}$$

Proceeding inductively as in § 4, we get

$$g_i^{(k)} \leq f_i^{(k-1)}, \qquad i = 1, \cdots, N - 1, \qquad k = 1, 2, \cdots \tag{2}$$

On the other hand,

$$g_i^{(1)} = \min_{j \neq i} (t_{ij} + g_j^{(0)}) \geq \min_{j \neq i} t_{ij} = F_i^{(0)} \tag{3}$$

Induction now yields

$$g_i^{(k)} \geq F_i^{(k-1)}, \qquad i = 1, \cdots, N - 1, \qquad k = 1, 2, \cdots. \tag{4}$$

Eqs. (2) and (4) show that the number $g_i^{(k)}$ lies between $F_i^{(k-1)}$ and $f_i^{(k-1)}$. Since $f_i^{(k-1)} = f_i$ and $F_i^{(k-1)} = F_i$ for k large enough, and since $f_i = F_i$ by uniqueness of the solution of (1.1), it follows that $g_i^{(k)} = f_i = F_i$ for k large enough. This proves that for an arbitrary initial approximation

$$g_i^{(0)} \geq 0, \qquad i = 1, 2, \cdots, N - 1, \qquad g_N^{(0)} = 0$$

the method of successive approximations produces the desired solution.

We have now proved conclusively that our method of solution of (1.1) is theoretically fool-proof. That is, no matter what initial approximation we use, as long as it is nonnegative, the iterates must converge in a finite number of steps to a set of numbers $\{f_i\}$. These numbers will comprise a solution of (1.1), and in fact the unique solution. Finally, this solution of (1.1) is the physically meaningful solution of the routing problem. Of course, as we have already discussed at some length in Chapter Three there can be serious practical obstacles to the computational use of our method.

11. Least Time to Move *k* Steps

In § 6, we introduced the "lower solution," and remarked that it corresponds to actual paths among the vertices. In this section we shall explain the nature of these paths.

We recall that the lower solution was defined by the choice*

$$
\begin{aligned}
F_i^{(0)} &= \min_{j \neq i} t_{ij} && (i = 1, 2, \cdots, N-1) \\
F_N^{(0)} &= 0
\end{aligned}
\tag{1}
$$

and the iteration equations $(k = 0, 1, 2, \cdots)$

$$
\begin{aligned}
F_i^{(k+1)} &= \min_{j \neq i} [t_{ij} + F_j^{(k)}] && (i = 1, 2, \cdots, N-1) \\
F_N^{(k+1)} &= 0
\end{aligned}
\tag{2}
$$

From (1) we see $F_i^{(0)}$ is the minimal time for a one-link path originating at vertex i $(i = 1, 2, \cdots, N-1)$; that is, for the path from i to that adjacent vertex which can be reached most quickly from i.

Let us now define $C_i^{(k)}$ to be the set of all paths originating at vertex i $(i = 1, 2, \cdots, N-1)$ and either

(a)   having exactly k links, and not having N as an internal vertex,† or else

(b)   terminating at vertex N after fewer than k links, and not having N as an internal vertex.

In these paths we allow circuits, including those from i to another vertex and back to i. Also we define $g_i^{(k-1)}$ to be the minimal time among all paths in class $C_i^{(k)}$, $(i = 1, 2, \cdots, N-1)$, $g_N^{(k-1)} = 0$, for $k = 1, 2, 3, \cdots$. Since $C_i^{(1)}$ is the set of one-link paths originating at i, it is clear that $g_i^{(0)} = F_i^{(0)}$.

---

* We could also have taken $\hat{F}_i^{(0)} = 0$ $(i = 1, 2, \cdots, N)$ and used Eq. (2). This would have led to the relations $\hat{F}_i^{(1)} = \min_{j \neq i} t_{ij} = F_i^{(0)}$ and $\hat{F}_i^{(k+1)} = F_i^{(k)}$. Therefore we can think of the numbers $F_i^{(k)}$ as arising from the "zero initial approximation," if we prefer.

† That is, not passing *through* vertex N.

We claim that the numbers $g_i^{(k)}$ satisfy the recurrence relations ($k = 0, 1, 2, \cdots$)

$$g_i^{(k+1)} = \min_{j \neq i} [t_{ij} + g_j^{(k)}] \qquad (i = 1, 2, \cdots, N-1)$$
$$g_N^{(k+1)} = 0 \tag{3}$$

An argument for this is as follows. Take $i \neq N$, $k \geq 0$, and let P be a path in class $C_i^{(k+2)}$ requiring the minimal time $g_i^{(k+1)}$. This path goes first from i to some vertex j different from i. If $j = N$, the path P ends at N and the time required is $t_{iN} = t_{iN} + g_N^{(k)}$. If $j \neq N$ the time required for P is t_{ij} plus the time for some path in $C_j^{(k+1)}$; in fact the path in $C_j^{(k+1)}$ must be the one of optimal time, $g_j^{(k)}$, since otherwise P would not be optimal. Therefore for some j we have

$$g_i^{(k+1)} = t_{ij} + g_j^{(k)}$$

Thus

$$g_i^{(k+1)} \geq \min_{j \neq i} [t_{ij} + g_j^{(k)}] \tag{4}$$

But if there is inequality in this relation, then for some vertex J

$$g_i^{(k+1)} > t_{iJ} + g_J^{(k)}$$

This would imply that there is a path which goes first from i to J, which is in class $C_i^{(k+2)}$, and which takes less time than P. This contradicts the definition of P. Therefore there must be equality in Eq. (4).

Comparing Eqs. (2) and (3), we see that the numbers $F_i^{(k)}$ and $g_i^{(k)}$ satisfy the same recurrence relations. We have already seen that $F_i^{(0)} = g_i^{(0)}$ ($i = 1, 2, \cdots, N$). It follows by induction that

$$F_i^{(k)} = g_i^{(k)} \qquad (i = 1, 2, \cdots, N; \; k = 0, 1, 2, \cdots) \tag{5}$$

Consequently, the interpretation of $g_i^{(k)}$ as the minimal time for paths in class $C_i^{(k+1)}$ applies to $F_i^{(k)}$. That is, *the iterate $F_i^{(k)}$ of the lower solution defined by (1) and (2) is the minimal time among all paths in class $C_i^{(k+1)}$* ($i = 1, 2, \cdots, N-1$; $k = 0, 1, 2, \cdots$).

This conclusion can be phrased a little differently. Suppose we let $\tilde{C}_i^{(k)}$ be the set of all paths originating at vertex i ($i = 1, 2, \cdots, N-1$) and either

(a)   having exactly k links, or else
(b)   terminating at vertex N after fewer than k links.

Such a path might pass through vertex N. However, if it does, then we can obtain a quicker path by stopping at N when we reach it; the resulting quicker path is still in class $\tilde{C}_i^{(k)}$, but also in class $C_i^{(k)}$. Consequently we see that a path which is optimal with respect to $\tilde{C}_i^{(k)}$, is also optimal with respect to $C_i^{(k)}$, or in other words, $F_i^{(k)}$ also is the minimal time among all paths in class $\tilde{C}_i^{(k+1)}$.

We leave it to the reader to show that $F_i^{(k)}$ is also the minimal time among all paths in $\hat{C}_i^{(k+1)}$, where $\hat{C}_i^{(k)}$ is the set of all paths originating at i and either

(a) having exactly k links, and not having N as in internal vertex, or else

(b) terminating at vertex N after fewer than k links, and containing no vertex more than once.

The arguments given here apply whether or not there is a link joining every pair of vertices. If not, then of course the minimization in (1), (2), and so forth is over vertices j accessible from i.

Exercises

1. Explain why $F_i^{(k)}$ is the minimal time among all paths in $\hat{C}_i^{(k+1)}$.

2. Consider the map in Fig. 2. The times for the various links are indicated on the figure. List all paths in $\tilde{C}_5^{(2)}$, $C_5^{(2)}$,

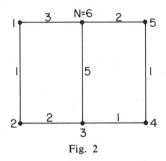

Fig. 2

and $\hat{C}_5^{(2)}$ and give the time for each. Compute $F_5^{(1)}$ from the recurrence relation and verify that it is the minimal time in each case.

3. Solve the quickest route problem for the map of Fig. 11 of Chapter One using the initial approximation $F_i^{(0)} = \min_{j \neq i} t_{ij}$. Compare this solution with those in §18 of Chapter Two.

4. Solve the Grand Canyon problem using the initial approximation $F_i^{(0)} = \min_{j \neq i} t_{ij}$.

12. Number of Iterations

In all the examples worked out previously, the iteration method has produced the exact solution in a relatively small number of steps. Naturally, it is of interest to know to what extent this will remain true for larger-scale problems. First of all, for the iteration scheme with initial policy

$$f_i^{(0)} = t_{iN} \qquad i = 1, \cdots, N - 1$$
$$f_N^{(0)} = 0 \qquad\qquad\qquad\qquad (1)$$

the interpretation of $f_i^{(k)}$ as the minimal time for a path from i to N with at most $k + 1$ links (k intermediate vertices) shows that $f_i^{(k)}$ must equal f_i for $k \geq N - 2$, since the optimal path (and any path without a closed circuit) can have no more than $N - 2$ intermediate vertices. Thus the iteration scheme using (1) is bound to yield the correct values of the f_i by the iterate $f_i^{(N-2)}$.

Next, consider the iteration scheme with initial policy

$$F_i^{(0)} = \min_{j \neq i} t_{ij} \qquad i = 1, \cdots, N \qquad (2)$$

Once again, the physical interpretation of the iterates is helpful. Recall that $F_i^{(k)}$ is now the minimal time in the class $\hat{C}_i^{(k+1)}$ of paths originating at i and either

(a) having $k + 1$ links, and not having N as an internal vertex, or else
(b) terminating at N in fewer that $k + 1$ steps, and containing no vertex more than once.

Let T be an upper bound on the time required for a path with at most $N - 1$ links and no closed circuits, terminating at N. For example, for T one could take the sum of the $N - 1$ largest values t_{ij} in the array of transition times. Since the optimal route from i to N can have no more than $N - 1$ links and can have no circuits, the optimal time f_i must satisfy the inequality

$$f_i \leq T$$

The optimal path from i to N, since it has at most $N - 1$ links, must be in the class $\hat{C}_i^{(N-1)}$, and in fact in $\hat{C}_i^{(k)}$ for $k \geq N - 1$. Consequently the opitimal time for a path in $\hat{C}_i^{(k)}$, which is $F_i^{(k-1)}$, cannot exceed f_i ($k \geq N - 1$). On the other hand, suppose that $F_i^{(k-1)}$ is less than f_i. Then there is a path in $\hat{C}_i^{(k)}$ with time $F_i^{(k-1)}$ which is quicker than the optimal path. This quicker path cannot reach N, since it would provide a route to N quicker than the optimal route, and it is therefore of type (a). That is, it has exactly k links. Now let the smallest transition time between any two vertices be

$$\tau = \min_{j \neq i} t_{ij} \qquad (3)$$

Then the path with k links requires a time of at least $k\tau$, and so its time $F_i^{(k-1)}$ is at least $k\tau$. It follows that

$$k\tau \leqq F_i^{(k-1)} < f_i \leqq T$$

This yields

$$k\tau < \frac{T}{\tau}, \qquad k \geqq N - 1 \tag{4}$$

For any k which satisfies $k \geqq T/\tau$ and $k \geqq N - 1$, it follows that the above situation cannot occur, or in other words $F_i^{(k-1)}$ cannot be different from f_i; hence, $F_i^{(k-1)} = f_i$. Since $T \geqq (N - 1)\tau$, the condition $k \geqq T/\tau$ implies $k \geqq N - 1$. Therefore, we can conclude that for the iteration scheme using (2), we must have

$$F_i^{(k)} = f_i \qquad i = 1, 2, \cdots, N \tag{5}$$

for $k \geqq (T/\tau) - 1$. The exact answer will thus be produced after at worst $(T/\tau) - 1$ iterations.

For the simple example of § 11, Exercise 2, one can take $N = 6$,

$$T = t_{36} + t_{16} + t_{23} + t_{56} + t_{21} = 13$$
$$\tau = 1$$

and see that the iteration process can take no more than 13 iterates (12 iterations). This bound is, of course, overly pessimistic, as can be verified by completing the iterative solution.

We proved in § 10 that if any nonnegative initial approximation $g_i^{(0)}$ is used, not necessarily (1) or (2), the successive approximation method still yields the correct values f_i. Moreover it is clear from § 10 that the number of iterates required can be no more than the number required for convergence of both the upper and lower iteration processes. Hence, it is no more than $T/\tau - 1$. Of course, since the initial guess (1) succeeds by the iterate $f_i^{(N-2)}$ at worst, one is not likely to use a different initial guess unless the physical situation suggests a good one.

Exercises

1. Find an upper limit to the number of iterations for the map in Fig. 11 of Chapter One. Compare with the actual number found in Exercise 3 of the last section.

2. Repeat Exercise 1 for the Grand Canyon problem.

Miscellaneous Exercises

1. Prove that in Dijkstra's algorithm (see Miscellaneous Exercise 1 in Chapter Two) the second vertex receiving a permanent label is the

vertex which can be reached most quickly from the starting vertex, and its label is the minimal time in which it can be reached. Prove that the vertex which can be reached in the next shortest time is the next vertex to be labelled permanently, and its label is the minimum time to reach it. Give a general proof that the nth vertex labelled permanently is the $(n-1)$st nearest (in time) to the starting vertex.

2. Consider the problem of cutting down on traffic congestion in the city. Suppose we want to accomplish this by having commuters drive their cars to a bus depot and then take a bus to the center of town. How would one go about determining the possible sites for a bus depot?

3. Suppose that several bus depots could be built. Where should they be placed? (N. B. These are not easy problems!)

4. *The Chifu-Chemulpo Puzzle.* A loop-line BGE connects two points B and E on a railway track AF, which is supposed blocked at both ends, as shown in Fig. 3. In the model, the track AF is 9 inches long, $AB = EF = 1^5/_6$ inches, and $AH = FK = BC = DE = {}^1/_4$ inch. On the track and loop are eight wagons, numbered successively 1 to 8, each 1 inch long and $^1/_4$ inch broad, and an engine, e, of the same dimensions. Originally the wagons are on the track from A to F and in the order 1, 2, 3, 4, 5, 6, 7, 8, and the engine is on the loop. The construction and the initial arrangement ensure that at any one time there cannot be more than eight vehicles on the track. Also if eight vehicles are on it, only the penultimate vehicle at either end can be moved on to the loop, but if less than eight are on the track, then the last two vehicles at either end can be moved on to the loop. If the points at each end of the loop-line are clear, it will hold four, but not more than four, vehicles. The object is to reverse the order of the wagons on the track so that from A to F they will be numbered successively 8 to 1: and to do this by means which will involve as

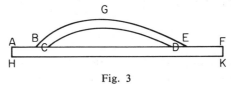

Fig. 3

few transferences of the engine or a wagon to or from the loop as is possible.

W. W. Rouse Ball, *Mathematical Recreations and Essays*, Macmillan, New York, (1947), 115.

5. *The Game of Bandy-Ball.* Bandy-ball, cambuc, or goff (the game so well known today by the name of golf), is of great antiquity, and

was a special favorite of Solvamhall Castle. Sir Hugh de Fortibus was himself a master of the game, and he once proposed this question.

They had nine holes, 300, 250, 200, 325, 275, 350, 225, 375, and 400 yards apart. If a man could always strike the ball in a perfectly straight line and send it exactly one of two distances, so that it would either go towards the hole, pass over it, or drop into it, what would the two distances be that would carry him in the least number of strokes round the whole course?

"Beshrew me," Sir Hugh would say, "if I know any who could do it in this perfect way; albeit, the point is a pretty one."

Two very good distances are 125 and 75, which carry you round in 28 strokes, but this is not the correct answer. Can the reader get round in fewer strokes with two other distances?

H. E. Dudeney, *The Canterbury Puzzles*, Dover Publications, Inc., New York, (1958) 58-59.

6. *The Tube Railway.* The following diagram is the plan of an underground railway. The fare is uniform for any distance, as long as you do not go along any portion of the line twice during the same journey. Now a certain passenger, with plently of time on his hands, goes daily from A to F. How many different routes are there from which he may select? For example, he can take the short direct route, A,

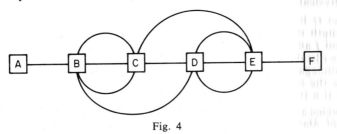

Fig. 4

B, C, D, E ,F, in a straight line; or he can go one of the long routes, such as A, B, D, C, B, C, E, D, E, F. It will be noted that he has optional lines between certain stations, and his selections of these lead to variations of the complete route. Many readers will find it a very perplexing little problem, though its conditions are so simple.

H. E. Dudeney, *The Canterbury Puzzles*, Dover Publications, Inc., New York, (1958) 149.

7. Consider a graph in which two positive numbers t_{ij} and s_{ij} are associated with the edge from i to j. The problem is to find the route from vertex 1 to vertex N for which the sum of the t_{ij} is least and the sum of the s_{ij} is greater than or equal to some fixed number, which we call x. Let

$$f_i(x) = \min \sum t_{ij}$$

among all paths (if any) such that

$$\sum s_{ij} \geqq x$$

Derive equations relating the $f_i(x)$ and discuss the question of existence and uniqueness of solutions and of computation of solutions. See

H. C. Joksch, "The Shortest Route Problem with Constraints," *J. Math. Anal. Appl.* **14** (1966) 191–197.

8.   Suppose that one defines an iteration

$$f_i^{(k)} = \min_{j \neq i} [t_{ij} + f_j^{(k-1)}], \qquad i = 1, 2, \cdots, N - 1$$
$$f_N^{(k)} = c_k$$

where $c_1, c_2, \cdots$, form a given sequence of numbers. Under what conditions on this sequence and on the initial approximation $f_i^{(0)}$ will the successive approximations converge in a finite number of iterations? What equations are satisfied by the limiting values f_i? See

D. L. Bentley and K. L. Cooke, "Convergence of Successive Approximations in the Shortest Route Problem," *J. Math. Anal. Appl.* **10** (1965) 269–274.

9.   Consider the iterative scheme

$$f_i^{(k)} = \min_{j \neq i} [t_{ij} + f_j^{(k-1)}] \qquad i = 1, 2, \cdots, N - 2$$
$$f_i^{(k)} = 0 \qquad\qquad\quad i = N - 1, N$$

$(k = 1, 2, 3, \cdots)$. Prove convergence of this iteration for any initial approximation $f_i^{(0)}$ $(i = 1, \cdots, N)$ for which $f_i^{(0)} \geqq 0$. What is the physical meaning of the set of limiting values?

10.   Let G be an undirected graph with n vertices. Let $d_{ij} = d_{ji}$ be the length of the shortest path (of one or more edges) between vertices i and j and let $d_{ii} = 0$. The symmetric matrix $D = (d_{ij})$ can be called the distance matrix of G. Now let D be a given symmetric matrix with nonnegative entries. Find a necessary and sufficient condition that D be the distance matrix of some graph. Is this graph unique? When is it a tree? See

S. L. Hakimi and S. S. Yau, "Distance Matrix of a Graph and Its Realizability," *Quart. Appl. Math.* **22** (1965) 305–317.

11.   Discuss the problem of finding a path $[v_0, v_1, \cdots, v_k]$ in a graph G which starts at vertex v_0, consists of a specified number K of edges, and is such that

$$\sum_{i=0}^{K-1} \alpha^i c_i$$

is a minimum, where c_i is the cost of the edge from v_i to v_{i+1} and where α is a given discount factor $(0 \leqq \alpha \leqq 1)$. The integer K can be interpreted as the number of time periods in the "planning horizon." See

J. F. Shapiro, "Shortest Route Methods for Finite State Space Deterministic Dynamic Programming Problems," *SIAM J. Appl. Math.* **16** (1968) 1232–1250.

Bibliography and Comment

For further discussion of theory and application, see

N. N. Moiseev, *Numerical Methods Using Variations in the Space of States*, Computing Center of the USSR Academy of Sciences, Moscow (1966) 1–22.

For the method of §4, see

D. L. Bentley and K. L. Cooke, "Convergence of Successive Approximations in the Shortest Route Problem," *J. Math. Anal. Appl.* **10** (1965) 269–274.

D. Davidson and D. J. White, "The Optimal Route Problem and the Method of Approximations in Policy Space," *J. Math. Anal. Appl.* (forthcoming).

Chapter Five

JUGGLING JUGS

1. A Classical Puzzle

Let us now turn our attention to a puzzle which has amused and be-mused people for hundreds of years.

Two men have a jug filled with eight quarts of wine which they wish to divide equally between them. They have two empty jugs with capacities of five and three quarts respectively. These jugs are unmarked and no other measuring device is available. How can they accomplish the equal division?

As usual, we shall approach the problem in a number of different ways. First, we shall make it precise. Then we shall discuss the trial-and-error approach. Next we shall consider a more systematic method of enumeration of possibilities. Following this, we shall show how the techniques developed in the previous chapters are applicable.

2. Precise Formulation

In stating the pouring problem, several assumptions were made tacitly. Let us see if we can make them explicit. First, we assume that no liquid is spilled (a conservation requirement), that none is lost due to evaporation, and that none is drunk (a prohibition requirement).

The fact that no measuring device is available means that the only operation that is allowed at any time is that of pouring wine from one jug to another. The pouring is stopped when one jug is empty, or the other is full, whichever event happens first.

Thus, for example, the first operation might be to pour five quarts from the full jug into the previously empty five-quart jug, or it might be to pour three quarts from the full jug into the previously empty three-quart jug. No other initial operation is possible.

3. Trial and Error

How can we grope our way toward a method for solving this riddle? One's first inclination, perhaps, is to adopt a method of trial and error, or experimentation. Of course, the experimentation need not actually be carried out physically, but can be carried out mentally, or indicated some-how on a piece of paper. For example, one could make a chart listing the amount of wine in each jug after each pouring operation. Thus, the chart below indicates that if we first pour from the first jug into the second, the jugs contain 3, 5, and 0 quarts, respectively; and if we then pour from the second into the third they contain 3, 2, and 3 quarts, respectively.   By means of such a chart one can keep track of a sequence of imaginary pour-ings. By trying various possibilities, one may perhaps eventually arrive at a solution to the puzzle, if indeed there *is* a solution.

TABLE 1

Number of pourings	8-quart jug	5-quart jug	3-quart jug
0	8	0	0
1	3	5	0
2	3	2	3

As a matter of fact, this puzzle has a number of solutions, three of which are given in Table 2.

Evidently this particular puzzle has various possible solutions.   It is natural to ask two questions:

(1)   Is there a systematic way of finding solutions in this problem or others of its kind?

(2)   How does one determine a "best" solution—the one requiring the fewest pourings, or the one requiring the least use of the three-quart jug, and so on?

4. Tree Diagrams

The discussion in the preceding section leads very naturally to a system-atic procedure for constructing solutions of the pouring puzzle. We need

TABLE 2

	Pourings	8-quart jug	5-quart jug	3-quart jug
First	0	8	0	0
solution	1	5	0	3
	2	5	3	0
	3	2	3	3
	4	0	5	3
	5	3	5	0
	6	3	2	3
	7	6	2	0
	8	6	0	2
	9	1	5	2
	10	1	4	3
	11	4	4	0
Second	0	8	0	0
solution	1	5	0	3
	2	5	3	0
	3	2	3	3
	4	2	5	1
	5	7	0	1
	6	7	1	0
	7	4	1	3
	8	4	4	0
Third	0	8	0	0
solution	1	3	5	0
	2	3	2	3
	3	6	2	0
	4	6	0	2
	5	1	5	2
	6	1	4	3
	7	4	4	0

only realize that whatever the contents of the various jugs are at a given stage, only certain pourings are possible—we can pour from one jug into another, stopping when there is no more to pour or when the latter jug is full. Thus, at any stage, only a few possibilities are present. If we can keep track of *all* possibilities, we shall surely find a solution eventually, if one exists.

To this end let us agree to represent the state of the jugs at any stage by three numbers, representing the number of quarts of wine in the eight,

five and three-quart jugs, respectively. Thus, (8, 0, 0) is the situation from
which we begin. The successive stages in the third solution of Table 1 are
then indicated by the list of triples:

$$(8, 0, 0)$$
$$(3, 5, 0)$$
$$(3, 2, 3)$$
$$(6, 2, 0)$$
$$(6, 0, 2)$$
$$(1, 5, 2)$$
$$(1, 4, 3)$$
$$(4, 4, 0)$$

In this symbolic way, (x, y, z) therefore indicates that there are x quarts in
the first jug, y quarts in the second, and z quarts in the third.

This kind of symbolism greatly simplifies the task of keeping track
of all possibilities which can arise. We can now say that starting from
(8, 0, 0) we either pour into the second jug until it is full, or into the
third jug until it is full, arriving at (3, 5, 0) or (5, 0, 3), respectively.
From (3, 5, 0) we can return to (8, 0, 0) or go to (0, 5, 3) or (3, 2, 3),
whereas from (5, 0, 3) we can go to (0, 5, 3), (8, 0, 0) or (5, 3, 0). We
want to continue listing all possibilites, and this is a good place to call
again on the technique of the *tree diagram*, introduced in Chapter One.
In fact, the diagram in Fig. 1 below contains all the information so far
given in this paragraph.

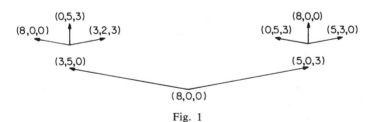

Fig. 1

Of course, there is clearly no sense in returning to (8, 0, 0) after
several pourings, and we can therefore cut out two of the branches from

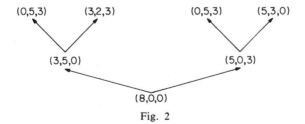

Fig. 2

Fig. 1, giving Fig. 2. In general, we can safely exclude an operation if it leads to a state which has already been encountered.

We can also alter the diagram by amalgamating identical states. Observe that (0, 5, 3) can be reached both from (3, 5, 0) and (5, 0, 3). Hence, we can replace Fig. 2 by the following.

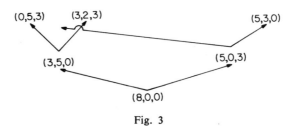

(0,5,3) (3,2,3) (5,3,0) (3,5,0) (5,0,3) (8,0,0)

Fig. 3

Observe that we use the standard circuit theory convention:

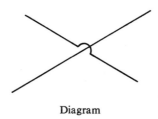

Diagram

This indicates that the paths do not intersect.

The procedure we follow to find a solution is now as follows. We construct the tree diagram, step by step. Each path through the tree from the root at (8, 0, 0) to the tip of a branch labelled (4, 4, 0) represents a solution of the puzzle. After any given number of pourings, the set of all possible paths through the tree represents the set of all possible pouring sequences (excluding returns to previous states). For example, Fig. 2 or Fig. 3 enables us to enumerate all possible sequences of two pourings:

$$(8, 0, 0) \rightarrow (3, 5, 0) \rightarrow (0, 5, 3)$$
$$(8, 0, 0) \rightarrow (3, 5, 0) \rightarrow (3, 2, 3)$$
$$(8, 0, 0) \rightarrow (5, 0, 3) \rightarrow (0, 5, 3)$$
$$(8, 0, 0) \rightarrow (5, 0, 3) \rightarrow (5, 3, 0)$$

Clearly there are exactly four ways to carry out two pourings. Also, there are three possible states for the jugs after two pourings. Finally, the totality of states which can be reached in either one or two pourings can be listed as follows:

$$(8, 0, 0), (3, 5, 0), (5, 0, 3), (0, 5, 3), (3, 2, 3), (5, 3, 0)$$

The similarity between this enumeration method and the one given in Chapter One is evident. There is, however, one significant difference. Namely, in the present case we cannot be sure in advance that the desired state (4, 4, 0) will *ever* be reached. Nor, at first glance, is it entirely clear that the tree diagram will eventually stop growing.

Exercises

1. Draw a complete tree diagram for the foregoing problem. Terminate a path with a *T* when it is impossible to continue without repeating.

2. Show from this diagram that there are several different ways to reach the state (4, 4, 0) from the state (8, 0, 0).

3. Show that *seven* is the minimun number of pourings needed to reach (4, 4, 0) and there is only one way to accomplish this.

4. Show that the tree does stop growing—in fact, it is not possible to pour more than 15 times without returning to a state already encountered.

5. Show that starting from (8, 0, 0), the only states which can be enconuntered are the following:

TABLE 3

(8, 0, 0)	(7, 1, 0)	(6, 2, 0)	(5, 3, 0)	(4, 4, 0)
	(7, 0, 1)	(6, 0, 2)	(5, 0, 3)	(4, 1, 3)
(3, 5, 0)	(2, 5, 1)	(1, 4, 3)	(0, 5, 3)	
(3, 2, 3)	(2, 3, 3)	(1, 5, 2)		

5. Some Further Observations

A number of rather surprising facts strike one when examining Table 3. For example, it is possible to end up with any number of quarts in any one of the jugs, within the stated capacity of the jug. On the other hand, not every conceivable arrangement of quantities in the three jugs is possible; for example, the triples (6, 1, 1), (5, 1, 2), and (5, 2, 1) do not occur.

We may well ask ourselves the reasons for the phenomena noted above. In one case, at least, these are readily supplied. It is obvious from the outset that the smallest jug can contain only 0, 1, 2, or 3 quarts at any stage, the next jug can contain only 0, 1, 2, 3, 4, or 5, and the largest jug 0, 1, 2, 3, 4, 5, 6, 7, or 8. Consequently, there are only a finite number (at the very most, $4 \times 6 \times 9 = 216$) of possible states for the jugs. If, therefore, we carry out a sequence of pourings, never returning to a state already encountered, we must eventually exhaust all the possibilities open to us. It follows that the tree diagram cannot grow indefinitely, but must terminate in a finite number of steps.

Some of the other points raised above will be discussed below, but first we shall discuss a helpful geometric representation of our problem, and the use of the digital computer in constructing the diagram, or enumerating cases.

Exercises

1. Draw the tree diagram if the three jugs contain 5, 3, and 2 quarts respectively, and initially the five-quart jug is full and the others empty.

2. Draw the tree diagram if the jugs have capacities 6, 3, aud 2 quarts respectively, and also if they have capacities 4, 3, and 2 quarts.

3. It it true that all links in the tree are two-way links? If we can go in one pouring from (x, y, z) to (x_1, y_1, z_1), can we always go back from (x_1, y_1, z_1) to (x, y, z) in one pouring? Can we return in any number of pourings?

6. Enumeration by Computer

We can attempt to construct a tree diagram (that is, to enumerate all possible sequences of operations) with the aid of a computer. A certain amount of thought is required to accomplish this because of the difficulty in explicitly setting forth the decision rules which the computer must follow in going from one case to the next. It is easy for us to see at a glance what operations are possible at each stage. It is however, very instructive to examine in detail the algebraic operations contained in each of these "glances." It is a frequent situation in the use of the computer that human perception is either very difficult, or impossible, to replicate.

Suppose, as usual, that we let

$x =$ amount of wine in first jug (8-quart jug)

$y =$ amount of wine in second jug (5-quart jug)

$z =$ amount of wine in third jug (3-quart jug)

Then the state of our system at a given stage is indicated by (x, y, z). For example, at the outset $x = 8$, $y = 0$, $z = 0$.

Suppose that the computer has generated part of the tree diagram and is ready to extend it. The computer must examine each state (x, y, z) which is currently the end of a branch and then determine what states can follow (x, y, z). A method for doing this is as follows. First, examine x. If $x > 0$ it is possible to pour from the first jug into one of the others. Therefore, y and z must be examined. If $y = 5$, the second jug is full and pouring into the third jug is possible. Thus if $x > 0$ and $y = 5$, the computer must go from (x, y, z) to $(0, y, 3)$. On the other hand (still in the case $x > 0$), if $y < 5$ and $z = 3$, pouring into the second jug is possible and the computer must go from (x, y, z) to $(0, 5, z)$. If $y < 5$ and $z < 3$, there are two possible branchings:

$$(x, y, z) \begin{cases} (x - 5 + y, 5, z) \\ (x - 3 + z, y, 3) \end{cases} .$$

The foregoing enumeration gives all possible pourings from the first jug into one of the others. The computer must also be programmed to generate all possible pourings from the second or third jug. Once this has been done, the computer can generate a list of all states attainable from the given state (x, y, z). These new states must then be added to the list of previous states in the branch and a test made to see whether they duplicate a previous state. If there is duplication, the new state is discarded. If not, the computer must move on to the terminal state of the next branch and repeat the process.

It is clear that to achieve all this a computer program can be quite complicated and its execution rather slow. We shall therefore pursue a quite different path below in obtaining our solution. The interested reader can try his hand at writing a program to generate all new states attainable from a given state (x, y, z) and to enumerate all possible sequences of pourings.

7. A Geometrical Representation

As we have seen, the number of quarts of liquid in the three jugs may conveniently be listed as a triple (x, y, z). This suggests the possibility of considering (x, y, z) as the rectangular coordinates of a point in a three-dimensional space, and this, of course, is one of the reasons why we used this notation. The set of all possible states then corresponds to a certain set of points. The coordinates x, y, z of these points are integers, none negative, and $x \leq 8$, $y \leq 5$, $z \leq 3$. Since the total amount of liquid in the jugs always remains eight quarts (a conservation relation), the relation $x + y + z = 8$ is satisfied; thus, all points (x, y, z) lie on the plane $x + y + z = 8$ in the positive octant. It is easy to list all the triples (x, y, z) satisfying the foregoing restrictions, as follows:

TABLE 4

(8, 0, 0)			
(7, 1, 0)	(7, 0, 1)		
(6, 2, 0)	(6, 1, 1)	(6, 0, 2)	
(5, 3, 0)	(5, 2, 1)	(5, 1, 2)	(5, 0, 3)
(4, 4, 0)	(4, 3, 1)	(4, 2, 2)	(4, 1, 3)
(3, 5, 0)	(3, 4, 1)	(3, 3, 2)	(3, 2, 3)
	(2, 5, 1)	(2, 4, 2)	(2, 3, 3)
		(1, 5, 2)	(1, 4, 3)
			(0, 5, 3)

There are 24 such triples: a few of the corresponding points are shown in Fig. 4.

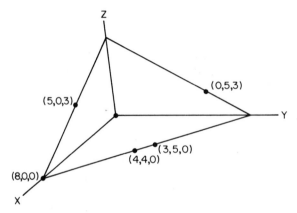

Fig. 4

A more convenient representation can be obtained if one observes that it is unnecessary to record all three of the numbers x, y and z, since by virtue of the relation $x + y + z = 8$, we can always find the third if we are given two of them. For example, we need only record x and y, and

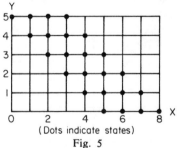

(Dots indicate states)

Fig. 5

remember that $z = 8 - x - y$. The various states are then represented by points (x, y) in the much more convenient coordinate plane, as shown in Fig. 5. Referring to Table 3, we notice the interesting fact that the states which can be reached from $(8, 0)$ all lie on the "outside edge" of the figure formed by the dots in Fig. 5, whereas the other states all lie "inside" the figure. The explanation of this will be given later.

If one records the pairs (y, z) rather than (x, y), Fig. 6 is obtained. Again the states which can be reached lie on the boundary of the rectangle of dots, and the others lie inside.

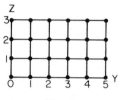

Fig. 6

A two-dimensional representation by pairs (x, y) or (y, z), rather than by triples (x, y, z), is also important from the computational point of view, since it reduces the amount of computer storage required.

Each time we pour from one container to another, the state triple (x, y, z) changes. This can be indicated geometrically by showing a path between the points corresponding to the old and new states. For example, the first six pourings in the first solution given in Table 2 can be shown on a yz-diagram as in Fig. 7. The third solution in Table 2 is shown in full in Fig. 8. The evidence so far indicates that the route in Fig. 8 is the shortest one from $(0, 0)$ to $(4, 0)$.

The geometrical pictures given here remind us somewhat of the maps in Chapter One. In fact, the pouring problem and the shortest-route problem are identical in structure. In each case, we have a finite number of

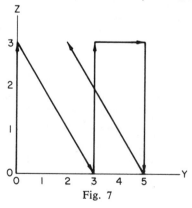

Fig. 7

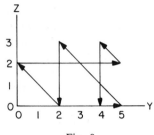

Fig. 8

points, from each of which transitions to certain other points are permissible. Each transition exacts a certain "cost"—the time t_{ij} for the routing problem and one operation for the pouring problem. The question in both cases is that of choosing the sequence of transitions to pass from a given starting point to a given terminal point which minimizes this cost.

There are, of course, certain differences between the two problems. The figures for the pouring problem do not, as in the routing problem, represent maps of physical terrain. Moreover, one cannot pass in one step from a given point to its adjacent points; rather one can move only to certain other points.* For example, from the state $(2, 3)$ one can go only to $(5, 3)$, $(5, 0)$, $(2, 0)$, or $(0, 3)$. The four points $(5, 3)$, $(5, 0)$, $(2, 0)$, $(0, 3)$ are the "nearest" points to $(2, 3)$, for they are the only ones which can be reached in a single pouring. One might say that the distance between points is measured in a noneuclidean fashion.

8. Cost Matrix

The geometrical interpretation of the pouring problem in the last section has shown us that the pouring problem and the shortest-route problem are of identical structure. Consequently, the techniques worked out for solutions of the latter problem must have counterparts for solutions of the former. Two of the principal ideas of Chapter One have already been carried over to the pouring problem, namely the reformulation of the puzzle in terms of transitions between states and the use of the tree-building or enumerative method. Another principal method of Chapter One, a reformulation in terms of functional equations, will now be applied to the pouring problem. The first step in doing this is to define quantities analogous to the t_{ij} and f_j of Chapter One. Instead of the time t_{ij} to move from one city to another, we shall use the "cost" of performing the transition from one state to another. There are various ways in which one can

* Think in terms of a Knight move in Chess. In German, this piece is called "Der Springer."

interpret the word "cost," the simplest of which is that each pouring operation costs one unit. In place of the array, or matrix, of times t_{ij}, we now have a matrix of costs.

Let us illustrate this for the pouring problem in which the containers hold 5, 3, and 2 quarts.* Using for example the result of Exercise 1 in §5, we find that the only possible states (x, y, z) which can be reached starting from $(5, 0, 0)$ are as follows:

$$(5, 0, 0) \quad (4, 1, 0) \quad (3, 2, 0) \quad (2, 3, 0) \quad (1, 3, 1) \quad (0, 3, 2)$$
$$(4, 0, 1) \quad (3, 0, 2) \quad (2, 1, 2) \quad (1, 2, 2)$$

A geometrical representation of states (y, z) is given in Fig. 9; as before, the 10 states which can be reached, starting from $(0, 0)$, lie on the outside of the rectangle of states which satisfy the requirements $0 \leq y \leq 3$, $0 \leq z \leq 2$.

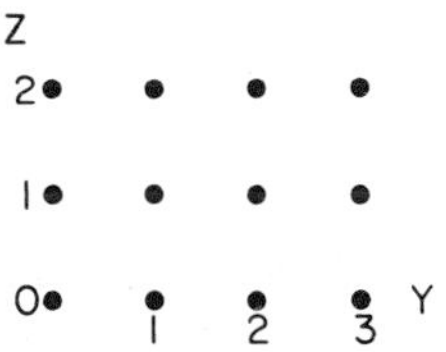

(Dots indicate states)

Fig. 9

TABLE 5

Cost Matrix (Containers 5, 3, 2)

To \ From	(0,0)	(0,1)	(0,2)	(1,0)	(1,1)	(1,2)	(2,0)	(2,1)	(2,2)	(3,0)	(3,1)	(3,2)
(0,0)			1							1		
(0,1)	1		1	1							1	
(0,2)	1						1					1
(1,0)	1	1				1				1		
(1,1)		1	1	1		1	1				1	
(1,2)			1	1						1		1
(2,0)	1		1						1	1		
(2,1)		1				1	1		1	1	1	
(2,2)			1				1				1	1
(3,0)	1					1						1
(3,1)		1							1	1		1
(3,2)			1							1		

* This matrix is used because it is smaller than the one for the problem with containers of size 8, 5, and 3.

We shall now construct the "cost matrix" for this example. Using the technique of Chapter One, we list each of the 12 possible states along the top and along the left side of a table. In the ith row and jth column of the array, we place the cost, one, for a transition from the state at left of the row to the state at the top of the column if such a pouring is possible. If no such transition is possible, no entry is made. In Table 5, we give the cost matrix for the example we are discussing; we leave it to the reader to verify the correctness of this table, and to discover such regularities as may exist in the result.

Exercise

1. How many states can be reached from each given starting state? Can you guess a rule which makes it possible to predict this number?

9. Connection Matrix

There is another array, closely related to the array of costs, which contains much of the information concerning transitions from one state to another. In this array, called the *connection matrix* (or *adjacency matrix*),[*] the element in a given row and column is one if the transition from the state indicated by the row to the state indicated by the column is possible in one step, and the element is zero if the transition is not possible in one step. In the present example, since the cost of a transition is in every case one, if it is possible at all, the connection matrix can be obtained from the cost matrix by inserting zeros in all the empty locations. The zeros in the diagonal locations, where the row and column numbers are the same, indicate that it is not possible to perform one pouring operation, and remain in the initial state. In other cases, the entries in the connection matrix will all be zeros and ones, but those in the cost matrix may by any appropriate numbers. In every case, zeros in the connection matrix occur in the same locations as blanks in the cost matrix.

Exercise

1. Construct the cost matrix for the 5-3-2 pouring problem if the cost in pouring is measured by the weight of the container lifted. You may assume that the jug has negligible weight, so that the weight lifted is proportional to the number of quarts of liquid in the jug.

---

[*] This definition agrees with that given in Chapter Two for the connection or adjacency matrix of a graph.

10. Minimum Cost Functions

Following the procedure of Chapter One, we now introduce a minimum cost function, as follows:

$f(x, y, z)$ = minimum cost at which it is possible to
reach the desired terminal state, starting
at state (x, y, z)

We shall immediately modify this definition, since, as we have seen, it is unnecessary to retain all three variables x, y, and z explicitly. Also, cost for the pouring problem will be assumed to be simply the number of pourings. Therefore we define

$f(y, z)$ = minimum number of pourings required to reach
the desired state, starting in state (y, z) (1)

There is a significant difficulty here which was not present in Chapter One. It is not apparent that the function $f(y, z)$ is always well-defined by (1). In fact, for the problem with jugs of sizes 5, 3, and 2, the state $(1, 1)$ cannot be reached, starting from the state $(0, 0)$ in any number of pourings. This means that $f(0, 0)$ is undefined if the desired state is $(1, 1)$.   Let us agree to define $f(y, z) = \infty$ whenever it is not possible to reach the desired state from (y, z) in a finite number of pourings.   This is similar to the stratagem employed in § 21 of Chapter Two.

11. Powers of the Adjacency Matrix

Let C denote the adjacency matrix, let $C^2 = C \times C$, the usual matrix product, let $C^3 = C^2 \times C$, and so on.   Then (see Chapter Two, § 20, Exercise 4) the entry in row i and column j of C^m represents the number of distinct ways of getting from state i to state j by a sequence of exactly m pourings $(m = 1, 2, 3, \cdots)$.   Suppose, then, that we look at the i, j entries in C, C^2, etc.   If these entries are 0 in $C, C^2, \cdots, C^{m-1}$ but the entry in C^m in nonzero, it follows that there is a way of getting from state i to state j in m pourings, but no way of doing this in fewer pourings.   Hence, m will be the minimal number of pourings to go from state i to state j.

We therefore see that we can find $f(y, z)$, for any starting state (y, z) and any desired state, by computing the successive powers C, C^2, C^3, and so on, and finding the smallest power (if any) for which the corresponding entry is nonzero.   We leave it to an Exercise for the reader to illustrate this.   In the next section, we shall pursue an alternative method, analogous to the functional equation method of Chapters One and Two.

Exercises

1. Let C be the adjacency matrix for the 5-3-2 pouring puzzle (see Table 5). By computing powers of C, find $f(y, z)$ for each (y, z) if $(0, 1)$ is the desired final state.

2. If C is an N by N matrix, how many operations are required to compute C^2? How many additional to compute C^3, C^4, etc? What can be said about the number of operations required to find the minimal number of pourings to go from state i to state j for all i and j?

12. Functional Equations

Our next step is to write functional equations relating the numbers $f(y, z)$, for various y and z, corresponding to Eq. (13.2) of Chapter One. The reasoning of Chapter One, § 13, is immediately applicable to the present situation, if we make the correct reinterpretation of the symbols. Indeed, Eq. (13.2) states that the minimum time to reach the desired final state, starting from state i, is the time t_{ij} to reach j plus the minimum time to reach the desired final state starting from j, where j is chosen optimally. If the word "time" is replaced by the word "cost," and t_{ij} by 1 (the cost of one pouring), this applies to the pouring problem. Consequently, we can write the following equation

$$f(y, z) = \min_{(u, v)} [1 + f(u, v)]$$

in which it is understood that the states (u, v) must be chosen from the set of all states accessible from (y, z) in one pouring. To emphasize that the state (u, v) must be different from (y, z), we may write

$$f(y, z) = \min_{(u, v) \neq (y, z)} [1 + f(u, v)] \tag{1}$$

Let us illustrate this by reference to Table 5. Since (0.2) and $(3, 0)$ are the only states which can be reached directly from $(0, 0)$, we have

$$f(0, 0) = \min [1 + f(0, 2), 1 + f(3, 0)]$$

The only states accessible from $(0, 2)$ are $(0, 0)$, $(2, 0)$, and $(3, 2)$. Thus

$$f(0, 2) = \min [1 + f(0, 0), 1 + f(2, 0), 1 + f(3, 2)]$$

Continuing in this way, we could write 12 equations, one for each possible state.

Eq. (12.1) is valid for each state (y, z), other than the desired final state, which we shall denote by (y_d, z_d). For it, we have the relation

$$f(y_d, z_d) = 0 \tag{2}$$

Incidentally, in case the final state (y_d, z_d) cannot be reached from (y, z), Eq. (12.1) remains valid, in the sense that $1 + f(u, v) = \infty$ for every (u, v) accessible in one step from (y, z); and consequently, the equation has the form $\infty = \infty$.

Exercise

1. Write the complete set of 12 functional equations corresponding to Table 5.

13. Successive Approximations

Eq. (12.1), combined with the method of successive approximations, provides us with an algorithm for constructing the minimal sequence of pourings to reach any desired final state, just as the analogous equations provided an efficient means of solving the shortest route problem (see Chapter Two). The basic equations for the successive approximations are now

$$f^{(k)}(y, z) = \min_{(u,\, v) \neq (y,\, z)} [1 + f^{(k-1)}(u, v)], \qquad k = 2, 3, \cdots,$$
$$f^{(k)}(y_d, z_d) = 0, \qquad k = 1, 2, \cdots, \tag{1}$$

with $f^{(1)}$ a known function. These equations are used in a by now familiar fashion. Let us illustrate this, using Table 5. As usual, there are various reasonable choices for the initial policy. We shall carry out the calculation for the choice

$$f^{(1)}(y, z) = \infty, \qquad (y, z) \neq (0, 1)$$
$$f^{(1)}(0, 1) = 0 \tag{2}$$

The desired state is $(0, 1)$. The calculation can be carried out using the paper strip method of Chapter Two. That is, the values of $f^{(1)}(y, z)$ are listed horizontally* on a paper strip in the same order as in Table 5, $(0, 0)$, $(0, 1)$, $\cdots$ $(3, 2)$. This strip is then placed in turn next to each row of Table 5. Wherever two numbers are side-by-side, they are added. The smallest of these sums becomes the new value for the state listed at the head of the row. The blank entries in Table 5 guarantee that the search for the minimum sum is never conducted among states which cannot be reached from a given state.

The result of this calculation is shown in Table 6. It is seen that the state $(0, 1)$ can be reached in at most four pourings from any starting state. For example, an optimal sequence of pourings beginning with five quarts in the largest jug may be indicated thus:

---

* Alternatively, list these vertically on a strip but use the transpose of the matrix in Table 5.

$$(0, 0) \to (3, 0) \to (1, 2) \to (1, 0) \to (0, 1) .$$

We leave it to the reader to discover what happens when this procedure is used if the desired final state cannot be reached from every possible starting position—see Exercise 1.

TABLE 6

SUCCESSIVE APPROXIMATIONS FOR THE POURING PROBLEM

(CONTAINERS 5, 3, 2)

State (y, z)	$f^{(1)}$ (y, z)	Next state	$f^{(2)}$ (y, z)	Next state	$f^{(3)}$ (y, z)	Next state	$f^{(4)}$ (y, z)	Next state	$f^{(5)}$ (y, z)
(0, 0)	∞		∞		∞		∞	(3, 0)	4
(0, 1)	0		0		0		0		0
(0, 2)	∞		∞		∞		∞	(2, 0)	4
(1, 0)	∞	(0, 1)	1	(0, 1)	1	(0, 1)	1	(0, 1)	1
(1, 1)	∞	(0, 1)	1	(0, 1)	1	(0, 1)	1	(0, 1)	1
(1, 2)	∞		∞	(1, 0)	2	(1, 0)	2	(1, 0)	2
(2, 0)	∞		∞		∞	(2, 2)	3	(2, 2)	3
(2, 1)	∞	(0, 1)	1	(0, 1)	1	(0, 1)	1	(0, 1)	1
(2, 2)	∞		∞	(3, 1)	2	(3, 1)	2	(3, 1)	2
(3, 0)	∞		∞		∞	(1, 2)	3	(1, 2)	3
(3, 1)	∞	(0, 1)	1	(0, 1)	1	(0, 1)	1	(0, 1)	1
(3, 2)	∞		∞		∞		∞	(3, 0)	4

Exercise

1. Carry through the successive approximation scheme using Table 5, and taking (1, 1) as the desired state. When can the calculation be terminated?

14. The Labelling Algorithm

At the end of Chapter Two, the method of succesive approximations with initial policy $f_i = \infty$ (except $f_N = 0$) was shown to correspond to certain "labelling algorithms." The same is true for the pouring problem, for which the labelling algorithm like that in § 24 of Chapter Two may be carried out as follows. First draw the state diagram, as in Fig. 9, labelling the desired terminal vertex with a zero and all other vertices with ∞. This is done in Fig. 10a, with (0, 1) the terminal vertex. At the next step, locate the set S_1 of vertices from which the vertex (0, 1) can be reached in one step, and change the label on each of these to one (cf., Fig. 10b). The location of the set S_1 is no longer an easy visual process, as it was in Chapter One, but is still possible with the aid of Table 5, the cost matrix. In fact, we merely look at the column corresponding to the state $(0, 1)$; the set S_1 consists of the states corresponding to rows in which there is an entry 1 in this column. Next, locate the set S_2 of states, not already included in S_1, from which a state in S_1 can be reached in one pouring. Label the vertices in S_2 with a 2 (Fig. 10c). Other labels can be left unchanged since all transitions have the same cost. In this respect, the algorithm is simpler than the one in § 24 of Chapter Two. Next, label with a 3 the states in S_3, from which a state in S_2 can be reached in one pouring, and so on. Comparison of Fig. 10 and Table 6 shows that this labelling method corresponds to the rise of successive approximations as explained above.

It follows from the discussion in § 22 of Chapter Two, and is evident from our example, that the successive iterates $f^{(k)}(y, z)$ have a simple physical meaning, when the initial policy is chosen as in (13.2). Indeed, $f^{(2)}(y, z)$ is 1 if the desired final state can be reached from (y, z) in one pouring, and otherwise is ∞; $f^{(3)}(y, z)$ is 1 if the desired final state can be

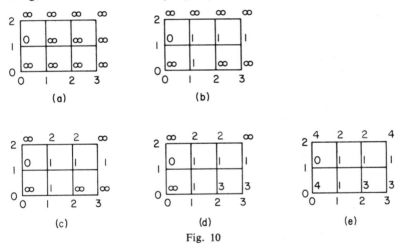

Fig. 10

reached in one pouring, 2 if it can be reached in two pourings but not in one, and ∞ otherwise. In general, if M is the minimal number of pourings in which the desired state can be reached from (y, z), then

$$f^{(k)}(y, z) = M \text{ for } k > M$$
$$= \infty \text{ for } k \leq M$$

Of course if $M = \infty$, then $f^{(k)}(y, z) = \infty$ for every k.

The considerations of Chapter Two, § 21, relating to "fictitious paths" and large initial times have their counterparts here, but we shall leave it to the reader to supply the details.

15.   Avoiding the Cost Matrix

For a pouring problem in which the three jugs have large capacities, there may be a very large number of possible states. For example, if the jugs hold 69, 37, and 32 quarts, respectively, then the states are represented by pairs (y, z) with $0 \leq y \leq 37$, $0 \leq z \leq 32$. The number of states is thus $38 \times 33 = 1254$. This means that the cost matrix has 1254 rows and 1254 columns, for a total of 1,572,516 entries. It is obvious that in such a situation, hand calculation is out of the question. Furthermore, we are straining the rapid access capabilities of even the largest of current computers.

In such a situation, it is essential to design a machine program which will avoid this difficulty. Three ways in which this can be done are as follows:

1.   Store the cost matrix on magnetic tapes or disc instead of in rapid-access storage (see Chapter Three, § 7).

2.   Do not store the cost matrix at all. Instead, store the accessibility lists (see Chapter Three, § 8). That is, for each state (y, z), store the set $S(y, z)$ of all states accessible from (y, z) in one pouring. When the connection matrix is sparse, this will greatly reduce the storage requirement.

3.   Do not store the cost matrix or the accessibility list. Instead, generate the accessibility lists as required. As suggested in § 9 of Chapter Three, we are thus replacing the storage of data by storage of an algorithm to generate the data.

Let us look in detail at the third alternative. Imagine that the numbers $f^{(k-1)}(y, z)$ have already been computed and that it is now desired to compute all the numbers $f^{(k)}(y, z)$. Pick a particular (y, z). Then it is necessary to apply Eq. (13.1), which means that we must find the set of all states (u, v) which can be reached in one pouring from the chosen (y, z), and look up the values $f^{(k-1)}(y, z)$. We see that what we need to do can be outlined thus:

(a)   Have the numbers $f^{(k-1)}(y, z)$ in storage, for all states (y, z).

(b)   Have a subroutine which generates all states (u, v) which can be reached in one pouring from (y, z) for each (y, z).

(c)   In the main program, call each (y, z), use (b) to find the states (u, v) and (a) to call up the corresponding values $f^{(k-1)}(u, v)$; perform the minimization required in Eq. (13.1), and store the resulting $f^{(k)}(y, z)$. When this has been done for every (y, z), replace* the stored values $f^{(k-1)}(y, z)$ by the values $f^{(k)}(y, z)$, and repeat the calculation.

---

* An alternative is to replace each $f^{(k-1)}(y, z)$ by $f^{(k)}(y, z)$ as soon as it is computed, rather than to wait until all are computed. This further reduces storage demands, and may be expected to speed up convergence as well; cf., the discussion in § 12 of Chapter Three.

This procedure reduces demands on rapid-access storage, since only the values $f^{(k-1)}(y, z)$ and $f^{(k)}(y, z)$, where y and z range over all possible sets, are stored. For the 69, 37, 32 pouring problem, for example, only $2 \times 1254 = 2508$ numbers need be stored. Of course, additional storage locations are needed in order to execute (b) and (c) above.

We leave it as an exercise for the reader to prepare a program along the lines of (a), (b), (c) if he wishes. In the next few sections we shall describe an alternate method which leads to somewhat different recurrence relations and programs.

16. Another Initial Policy

If the method of successive approximations is defined as in Eq. (13.1), but the initial policy is taken to be

$$f^{(1)}(y, z) = 0, \text{ all } (y, z) \tag{1}$$

instead of that in Eq. (13.2), a different sequence is generated. We leave it to the reader to compute this sequence for 5-3-2 pouring problem. Now $f^{(k)}(y, z)$ is a nondecreasing sequence, for each fixed (y, z). Moreover,

$$f^{(k)}(y, z) \leq k - 1$$

and $f^{(k)}(y, z) = k - 1$ if the desired final state cannot be reached from (y, z) in $k - 1$ or fewer pourings; otherwise, $f^{(k)}(y, z)$ is the least number of pourings in which the desired state can be reached from $(y\ z)$.

Exercise

1. Compute the successive approximations for the 5-3-2 pouring puzzle using the initial approximation (1).

17. Getting Close

In the wine pouring problem (as well as in life in general), it is not always possible to attain a desired objective. For example, in the problem with jugs of capacities 5, 3, and 2, it is impossible to reach the state in which there are, respectively, 3, 1, and 1 quarts in the three jugs, starting from any other state whatsoever. It is a natural human tendency, when a goal is unattainable, to settle for a little less—or to decide that the original goal was not so desirable after all. Moreover, in many situations, it is not particularly important that a desired state be precisely attained, but it is extremely important that it be closely approached. For example, it is probably not essential that a rocket sent to rendezvous with the planet Mars achieve exactly a prescribed distance from Mars, but only that it pass within an allowable distance within which meaningful observations can be made.

Thus, in the pouring problem, when a specified state is unattainable, we might attempt to get as *close* to the desired state as possible. But what do we mean by "close" here? That is, what can be meant by the "distance" between two states?

One interpretation of the meaning of "distance between two states" is suggested by the geometrical representation of the states, as given in § 7. We can simply measure the ordinary geometrical distance between the points corresponding to the two states. Thus if the states are (x_1, y_1, z_1) and (x_2, y_2, z_2), the distance between them would be

$$[(x_2 - x_1)^2 + (y_2 - y_1)^2 + (z_2 - z_1)^2]^{1/2} \qquad (1)$$

For example, the in 5, 3, 2 problem, the state $(3, 1, 1)$ is at a distance of $\sqrt{6}$ from the initial state $(5, 0, 0)$, at a distance of $\sqrt{8}$ from $(1, 3, 1)$, and so on. The "closest" states to $(3, 1, 1)$ are the states $(4, 0, 1)$, $(3, 0, 2)$, $(4, 1, 0)$, $(2, 1, 2)$, $(3, 2, 0)$, and $(2, 2, 1)$, all of which are at a distance of $\sqrt{2}$.

Since our geometrical representation of the pouring problem is our own creation, we need not feel obliged to use the geometrical distance as a measure of how close two states are to one another, and indeed other definitions are certainly feasible. Another possibility is to replace the expression in Eq. (17.1) by

$$|x_2 - x_1| + |y_2 - y_1| + |z_2 - z_1| \qquad (2)$$

This is a satisfactory measure, because it is zero if the two states coincide, and positive if they do not. With this definition, the point $(3, 1, 1)$ is 4 units from both $(5, 0, 0)$ and $(1, 3, 1)$; the closest points to $(3, 1, 1)$ are the same as before, but the distance is now 2 units.

The reader may well feel that the definition of "distance" in Eqs. (17.1) or (17.2) possesses no physical meaning and is thus rather pointless. The right way to consider it is as a device which enables us to replace Eq. (12.1), which involves the unknown function on both sides of the equations, with an equation, such as (18.2) below, which allows a direct iterative resolution, and is thus immediately amenable to the digital computer.

A reasonable excuse for introducing a device of this nature is that this particular stratagem is standard operating procedure in many parts of advanced mathematics. We try to pursue an intuitive and logical path as far as possible. Ocassionally, however, it is necessary to resort to ingenuity to circumvent a major obstacle to progress. It is this resort to wit which distinguishes mathematics from logic, or, more precisely, relegates logic to a relatively minor place in the mathematical hierarchy.

18.   Making the Closest Approach

The introduction of the idea of distance between two states leads us to the following questions:
   (1)   How can the reachable states closest to the desired state be determined? How close are they to the desired state?
   (2)   How is it possible to reach these these closest points? How long does it take to reach them—that is, what is the minimal number of pourings?

Once again, these questions can be answered by making a complete enumeration or tree, starting with a given initial state. Let us now show that it is also possible to utilize a simple recurrence relation. Let us use the symbol $d(P, Q)$ to represent the distance—in whatever sense is selected —between the states P and Q. Define $f_N(y, z)$ to be the minimum distance from the desired state which can be attained in N or fewer pourings, starting from the state (y, z). For example, starting from $(5, 0, 0)$ it is possible to reach either $(2, 3, 0)$ or $(3, 0, 2)$ in one pouring. Using the function in Eq. (17.2) as the notion of distance, these are at distances 4 and 2, respectively, from the terminal state $(3, 1, 1)$, and $(5, 0, 0)$ is at distance 4. Therefore $f_1(0, 0) = 2$.

The definition of $f_N(y, z)$ can be written in symbols as follows:

$$f_N(y, z) = \min_Q d(P, Q) \tag{1}$$

where P is the desired final state, and the minimum is sought among all states Q which can be reached in N or fewer pourings starting from (y, z). It is evident that $f_N(y, z)$ exists for all y, z, and N, since, once again, there are only a finite number of possible states. In those cases in which $f_N(y, z)$ is zero, the desired state P can be occupied in N or fewer pourings, starting from (y, z).

As in previous situations, we can write down a functional equation satisfied by the function $f_N(y, z)$ as follows:

$$f_N(y, z) = \min_{(u, v)} f_{N-1}(u, v), \qquad N \geq 1 \tag{2}$$

the minimum being taken over all states in the set S_1^* of states which can be reached from (y, z) in one pouring, or in no pouring at all. Note that in Eq. (18.2) we do not have the restriction $(u, v) \neq (y, z)$, because we have defined $f_N(y, z)$ to be the minimum distance attainable in N or *fewer* pourings.

A significant point about introducing $f_N(y, z)$ as the basic unknown, instead of $f(y, z)$ as previously defined, is that it provides more information. Not only does it indicate whether the desired state can be achieved, but also how close to it one can approach in a given number of steps if it cannot be achieved.

19. Calculation of $f_N(y, z)$

Either Eqs. (18.1) or (18.2) can serve as the basis of a method for computing the quantities $f_N(y, z)$. First let us consider Eq. (18.1). To employ this equation, one can find the successive sets of states S_1^*, S_2^*, $\cdots$, S_N^*, where S_1^* is the set of states which can be reached from (y, z) in one pouring or less; S_2^* is the set of states which can be reached in zero, one, or two pourings; and so on. This procedure is similar to that used in the labelling algorithm (§ 14), and can be carried out by reference to the connection matrix. Once the set S_N^* is found, $f_N(y, z)$ may be found by computing $d(P, Q)$ for every Q in S_N^*, and picking the minimum (P is the desired state).

For the 5-3-2 pouring problem, use of Table 5 leads to the following results, if we take $(0, 1)$ as the state P, and $(y, z) = (0, 2)$:

S_1^* contains the states   (0, 2),   (0, 0),   (2, 0),   (3, 2)
S_2^* contains the states   (0, 2),   (0, 0),   (2, 0),   (3, 2),
                          (3, 0),   (2, 2),
S_3^* contains the states   (0, 2),   (0, 0),   (2, 0),   (3, 2),
                          (3, 0),   (2, 2),   (1, 2),   (3, 1)

The distance of the states in S_1^* from the state $(0, 1)$, as measured according to Eq. (17.2), are respectivey 2, 2, 4, 4; hence, $f_1(0, 2) = 2$. The distances of states in S_2^* are 2, 2, 4, 4, 6, 6; hence, $f_2(0, 2) = 2$. It turns out that also $f_3(0, 2) = 2$, but $f_4(0, 2) = 0$, since S_4^* contains the state $(0, 1)$.

Let us formulate a computer algorithm for this procedure. We begin by numbering the states, and for each one, compute its distance from the desired terminal state. These distances are stored in the computer. Now, let an initial state s_0 be given. Its distance from the terminal state is examined; if zero, the program halts, otherwise we record the distance, $f_0(s_0)$, and go to the next step. This step is to generate the set S_1 of all states which can be reached in one step from the initial state. This can be done by use of a suitable subroutine.

We let $S_1^* = S_1 \cup \{s_0\}$. For each state in S_1, the distance to the terminal vertex is examined. If any is zero, the process terminates; otherwise the minimum among these distances and $f_0(s_0)$ is computed and recorded as $f_1(s_0)$. At the next step, for each state in S_1 the subroutine generates the set of states reachable in one step. All new states—that is, states not in S_1^*—are now added to S_1^* to form S_2^*, and the distance of each from the terminal vertex is examined. This procedure continues until a zero distance is found, or until a pre-assigned number of steps has been completed. The print-out should include the minimal distance found, the state at minimal distance and the path to it, and the number of steps.

Because of the repeated need to enlarge the sets S_k^* to S_{k+1}^*, by adding an undetermined number of new states, it is a fairly complicated matter to keep all the lists correct. Indeed, this procedure is really just that of forming the tree growing from the initial state, which (as we saw in Chapter One) is a lengthy process.

We shall now explain a method of calculation based on Eq. (18.2). First of all, $f_0(y, z)$ is the distance of (y, z) itself from the desired state P. Let this be computed for every state (y, z), and stored. According to Eq. (18.2),

$$f_1(y, z) = \min_{(u, v)} f_0(u, v)$$

where the minimum is taken over all states (u, v) in the set of states which can be reached from (y, z) in zero or one pourings. This set of states can be read off the connection matrix, or computed using a subroutine as suggested above. The values $f_0(u, v)$ have already been computed. Therefore $f_1(y, z)$ can be found. If $f_1(y, z)$ is found for every (y, z), this procedure can be repeated to find $f_2(y, z)$ for every (y, z), and so on.

This calculation can be conveniently set down in a paper strip algorithm, which we shall illustrate for the 5-3-2 pouring problem. First, compute the numbers $f_0(y, z)$ for all (y, z), and write these on a horizontal strip. Now form a modified cost matrix, identical to the cost matrix except that a 1 is placed in each diagonal location; this will indicate that in any state it is possible to remain in the state by performing no pouring operation. Suppose the f_0 strip is placed next to a row of the matrix, say the row headed $(0, 0)$. Each one in this row indicates that a transition is possible from $(0, 0)$ to the state in its column, in one or no pourings. Thus, the value $f_1(0, 0)$ is the minimum value of elements on the horizontal strip which appear next to 1's in the $(0, 0)$ row. Similarly $f_1(y, z)$ can be found for each (y, z). Then the numbers $f_1(y, z)$ can be written on a horizontal strip. If this strip is placed alongside a row of the modified cost matrix, say the row headed (a, b), then $f_2(a, b)$ is the minimum of the values on the strip which appear next to 1's in the (a, b) row.

The results of this calculation are given in Table 7. They provide the minimum distance, as defined by Eq. (17.2), from the assigned terminal state $(0, 1)$, for every initial state, which can be achieved in 1, 2, 3, etc., pourings. It appears from Table 7 that $f_4(y, z) = 0$ for every (y, z), which means that the terminal state can be reached in at most four pourings from any initial configuration. From the state $(3, 2)$, a path to the terminal state can be seen from the next to the last column to go first to $(3, 0)$. This reduces the distance from 8 to 6. From $(3, 0)$, the best policy is to go to $(1, 2)$ which reduces the distance to 4, then to $(1, 0)$. This path requires four pourings.

The method based on Eq. (18.2) is seen to provide a great deal of information, and indeed a solution for every initial state. Moreover, it seems to be easier to adapt to machine calculation than the one based on Eq. (18.1). Its only defect is that the calculation must be carried through for every state.

TABLE 7

CALCULATION OF $f_N(x, y)$ FOR THE POURING PROBLEM
(CONTAINERS 5, 3, 2) USING EQ. (18.2)

TERMINAL STATE: $(0, 1)$

State (y, z)	$f_0(y, z)$	Min state	$f_1(y, z)$	Min state	$f_2(y, z)$	Min state	$f_3(y, z)$	Min state	$f_4(y, z)$
(0, 0)	2	(0, 0)	2	(0, 0)	2	(0, 0)	2	(3, 0)	0
(0, 1)	0	(0, 1)	0	(0, 1)	0	(0, 1)	0	(0, 1)	0
(0, 2)	2	(0, 2)	2	(0, 2)	2	(0, 2)	2	(2, 0)	0
(1, 0)	2	(0, 1)	0	(0, 1)	0	(0, 1)	0	(0, 1)	0
(1, 1)	2	(0, 1)	0	(0, 1)	0	(0, 1)	0	(0, 1)	0
(1, 2)	4	(0, 2)	2	(1, 0)	0	(1, 0)	0	(1, 0)	0
(2, 0)	4	(0, 0)	2	(0, 0)	2	(2, 2)	0	(2, 2)	0
(2, 1)	4	(0, 1)	0	(0, 1)	0	(0, 1)	0	(0, 1)	0
(2, 2)	6	(0, 2)	2	(3, 1)	0	(3, 1)	0	(3, 1)	0
(3, 0)	6	(0, 0)	2	(0, 0)	2	(1, 2)	0	(1, 2)	0
(3, 1)	6	(0, 1)	0	(0, 1)	0	(0, 1)	0	(0, 1)	0
(3, 2)	8	(0, 2)	2	(0, 2)	2	(0, 2)	2	(3, 0)	0

Exercises

1. Write flow charts for the foregoing methods.

2. Use the paper strip method to analyze the 8-5-3 pouring puzzle. Find the minimal number of pourings required to go from each possible initial state to the terminal state $x = 4$, $y = 4$, $z = 0$.

3. Write a computer program for the method based on Eqs. (18.2). Apply this program to the 29-17-12 pouring puzzle to find the minimal number of pourings required to achieve the state $x = 28$, $y = 0$, $z = 1$, starting from each possible initial state $x = a$, $y = b$, $z = c$, $a + b + c = 29$.

4. Modify the previous program so that it is not necessary to store the cost matrix. Apply the new program to the 69-37-32 pouring puzzle.

20. Accessible and Inaccessible States

In Table 3 above, we listed all states for the 8-5-3 pouring problem which can be reached from the initial state $(8, 0, 0)$. In Fig. 5 and 6, these states lie on the outside boundary of the region of possible states. Points inside the region correspond to states which can never be reached from $(8, 0, 0)$.

The reason why inside points can never be reached is easily seen. Each pouring operation stops either when the jug from which we pour is empty, or the one into which we pour is full. In either case, at the conclusion of the pouring there must be at least one jug which is either full or empty. That is, if n_e denotes the number of empty jugs and n_f the number of full jugs, then is always true that

$$n_f + n_e \geq 1 \tag{1}$$

In Fig. 6, the inside points represent states in which the two smaller jugs are neither full nor empty. Moreover, the large jug is full if and only if the two smaller ones are empty, and it is empty if and only if the smaller ones are full. Consequently, no inside point in Fig. 6 can correspond to a state in which any of the three jugs is either full or empty.

It follows that the relation in Eq. (20.1) is violated at every inside point, and therefore that the state of the system after one or more pourings must correspond to a point on the boundary in Fig. 6. If the initial state corresponds to an interior point, the next and all subsequent states will correspond to boundary points.

The argument just given is valid for the general pouring problem with containers of capacities a, b, c respectively $(a = b + c, b > c > 0)$. If we let x, y, z denote the amounts in the three containers, as usual, then the possible states (y, z) satisfy $0 \leq y \leq b$, $0 \leq z \leq c$, and are represented by a rectangular region of dots in the (y, z) plane. At inside points in the diagram, one has $1 \leq y \leq b - 1$, $1 \leq z \leq c - 1$, and therefore using $x = b + c - y - z$, also $2 \leq x \leq b + c - 2$. Consequently no interior point can correspond to a state in which the relation in Eq. (20.1) is valid. We conclude that: *For the pouring problem with containers of capacities a, b, c $(a = b + c, b > c > 0)$, the state after one or more pourings corresponds to a boundary point on the state diagram.*

The states corresponding to interior points of the state diagram will be called *inaccessible*, since none of them can be reached starting from any other state. On the other hand, a state will be called *possibly accessible* if it corresponds to a point on the boundary of the (y, z) state diagram. It will be called *completely accessible* or *completely reachable* if it can be reached from every other state in a finite number of steps. In the next chapter we shall discuss a method for determining which states on the boundary of the state diagram are completely accessible.

Exercises

1. For the pouring problem with containers of capacities 6, 5, and 3, draw the (y, z) state diagram. Which states are inaccessible?

2. Consider the pouring problem with containers of capacities a, b, c $(b + c > a > b > c > 0)$, the first initially full. What is the shape of the region of possible states on the (y, z) state diagram? Prove that every interior point is inaccessible.

3. Repeat Exercise 2 if $a > b + c > b > c$.

Miscellaneous Exercises

Preliminary: Every so often, the "Chinese Fifteen" puzzle epidemic erupts. All over the country, millions of people are busily shuttling little squares over a 4 × 4 board trying to obtain certain preferred arrangements. The initial arrangement may be

1	2	3	4
5	6	7	8
9	10	11	12
13	14	15	X

Fig. 11

and the desired arrangement may be almost the same except that the 14 and 15 are interchanged.

At each stage, one can slide a number in a horizontal or vertical fashion into the blank square. Thus, at the end of the first move the possible positions are

1	2	3	4
5	6	7	8
9	10	11	X
13	14	15	12

1	2	3	4
5	6	7	8
9	10	11	12
13	14	X	15

Fig. 12

It may, for example, be desired to obtain in this fashion the boustrophedonic* position

* Derived from the path pursued by a bull plowing.

1	2	3	4
8	7	6	5
9	10	11	12
X	15	14	13

Fig. 13

In what follows, we shall consider some questions connected with the puzzle.

1. Consider the 2 × 2 case where the initial position is

1	2
X	3

Fig. 14

Show by direct enumeration that not all possible arrangements can be obtained starting from the foregoing.

2. Is there any initial position from which one can reach all other positions?

3. How many different positions can one reach from the foregoing positions? How many different closed chains are there?

4. Consider the 3 × 3 case where the initial position is

1	2	3
4	5	6
7	8	X

Fig. 15

Use the approach of the preceding pages to determine the minimum number of transitions required to attain a given position.

5. Are all positions attainable? Formulate a problem of attaining a position "nearest" to a given unattainable position.

6. Can you derive a simple rule for predicting when a position is attainable or not?

7. Is the 4 × 4 case soluble with a digital computer of a rapid access storage of 15!, of 13!, of 10!?

8. Consider an array of five numbers and a blank space. Can one obtain

Fig. 16

the following configuration?

Fig. 17

Treat this problem by considering the minimum distance that can be attained, and a terminal configuration where the distance is meas-

Fig. 18

ured by $|a_1 - 1| + |a_2 - 2| + \cdots + |a_5 - 5|$.

If the distance is zero, determine the minimum number of moves required to reach the given configuration. For a detailed discussion of this puzzle, see

W. W. Rouse Ball, *Mathematical Recreations and Essays*, Macmillan, New York, 1947, 299.

Preliminary: Lewis Carroll invented the game of "Doublets." The idea of the game is to construct a chain of English words connecting two given words subject to the condition that each word differs from the preceding word by the change of exactly one letter.

9. Show that LASS can be converted into MALE in four steps.

10. Show that WINTER can be converted into SUMMER.
    For the foregoing, see

M. Gardner, *New Mathematical Recreations from Scientific American*, Simon and Schuster, New York, 1966.

11. Consider the set S of all admissible English words of M letters. Let p be an element in this set and consider the set of transformations $T(p)$ which convert p into another element in the set by the change of a single letter. Let P_N be a given element in the set and let

$$f(p) = \text{the minimum number of transformations required to go from } p \text{ to } P_N$$

If there is no path, set $f(p) = \infty$. Show that

$$f(p) = 1 + \min_{p_i \in S} f(p_i)$$

12. Discuss the feasibility of computational solution of this problem for $M = 2, 3, 4, \cdots$

13. Consider the problem of best fit to P_N in a prescribed number of transformations. (Considering Lewis Carroll's affection for the word "fit," it is only fitting to use it here.)

    This game is related to the "word equivalence problem" in the theory of associative calculi. There one considers words in an arbitrary alphabet and a set of admissible substitutions and poses this problem: given any two words in the calculus, determine whether or not they can be connected by a chain of substitutions. This problem is algorithmically unsolvable, as pointed out in

B. A. Traktenbrot, *Algorithms and Automatic Computing Machines*, D. C. Heath, Boston, 1963.

For connections with genetics, see

J. M. Smith, *The Limitations of Molecular Evolution*, *The Scientist Speculates*, I. J. Good (Ed.) Basic Books, New York, 1962, 252–256.

For connections with philology and restoration of documents, see

A. Kaufmann and R. Faure, *Introduction to Operations Research*, Academic Press, 1968.

The foregoing is taken from

R. Bellman, "Dynamic Programming and Lewis Carroll's Game of Doublets," *Bull. Inst. Math and Its Applications* (forthcoming).

14. Starting in state (x, y), what is the maximum number of pourings we can make without ever returning to a previous state?

15. Three jars contain 19, 13, and 7 quarts, respectively. The first is empty, the others full. How can one measure out 10 quarts, using no other vessels?

M. Kraitchik, *Mathematical Recreations*, Dover Publications, Inc., New York, 1953, 30.

16. Three persons are to divide among themselves 21 equal casks, of which 7 are full, 7 are half full, and 7 are empty. How can an equable division be made without pouring the wine from any casks into others in such a way that each person receives the same amount of wine and the same number of whole casks?

Ibid.

17. A farmer leaves 45 casks of wine, of which 9 each are full, three-quarters full, half full, one-quarter full, and empty. His five nephews

want to divide the wine and the casks without changing wine from cask to cask in such a way that each receives the same amount of wine and the same number of casks. Furthermore, each shall receive at least one of each kind of cask, and no two of them shall receive the same number of every kind of cask.

Ibid., p. 31.

18. Two players A and B have three separate piles of coins on a table before them, each pile having a random number of coins. Each player in turn removes one or more coins from *one only* of the piles (he can remove one complete pile, if he wants to) with the object of finally forcing his opponent to remove the last single coin. How can you play this game to make certain of winning?

L. S. Graham, *Ingenious Mathematical Problems and Methods*, Dover Publications, Inc., New York, 1959, 10.

19. Using a set of balances, what is the least number of weighings necessary to discover which one of 12 coins of given denomination is counterfeit, assuming that the 11 standard coins are of equal weight and that the counterfeit coin is either lighter or heavier than a standard coin?

20. "Every child knows how to play this game. You make a square of nine cells, and each of the two players, playing alternately, puts his mark (a nought or a cross, as the case may be) in a cell with the object of getting three in a line. Whichever player first gets three in a line wins with the exulting cry:

> Tit, tat, toe,
> My last go;
> Three jolly butcher boys
> All in a row.

It is a very ancient game. But if the two players have a perfect knowledge of it, one of three things must always happen. (1) The first player should win; (2) the first player should lose; or (3) the game should always be drawn. What is correct?"

H. E. Dudeney, *The Canterbury Puzzles*, Dover Publications, Inc., New York, 1958, 156.

21. "Having examined 'Noughts and Crosses,' we will now consider an extension of the game that is distinctly mentioned in the words of Ovid. It is, in fact, the parent of 'Nine Men's Morris,' referred to by Shakespeare in *A Midsummer Night's Dream* (Act ii, Scene 2). Each player has three counters, which they play alternately on to the nine points shown in the diagram [Fig. 20], with the object of getting three in a line and so winning. But after the six counters are played they

then proceed to move (always to an adjacent unoccupied point) with the same object. In the example below White played first, and Black

Fig. 19

has just played on point 7. It is now White's move, and he will undoubtedly play from 8 to 9, and then, whatever Black may do, he will continue with 5 to 6, and so win. That is the simple game. Now, if both players are equally perfect at the game what should happen? Should the first player always win? Or should the second player win? Or should every game be a draw? One only of these things should always occur. Which is it?"

Ibid., pp. 156–157.

Bibliography and Comment

The pouring puzzles and the puzzles in the Miscellaneous Exercises are discussed in a number of books. In addition to those previously cited, see

M. Gardner, *Mathematical Puzzles and Diversions*, Simon and Schuster, New York, 1961.

N. A. Court, *Mathematics in Fun and In Earnest*, Signet Science Library, New American Library, New York, 1961.

T. H. O'Beirne, *Puzzles and Paradoxes*, Oxford University Press, New York and London, 1965.

E. Lucas, Récréations Mathématiques, deuxième édition, A. Blanchard, Paris, 1960.

Chapter   Six

THE SAWYER GRAPH AND THE
BILLIARD BALL COMPUTER

1. The Sawyer Graph

In the preceding chapter, we have given a precise formulation of the
wine-pouring puzzles and explained how our ideas of tree diagrams,
graphs, functional equations, and successive approximations can be adapted to yield systematic methods for resolving these puzzles. The purpose
of this chapter is to obtain a number of theoretical results regarding the
existence of minimal cost solutions and the classification of states as inaccessible, completely accessible, or possibly accessible. In order to achieve
this we shall draw graphs and state diagrams again and reason carefully
about the structure of these diagrams. In addition, we shall describe a
convenient graphical method for solving these puzzles, at least in the simpler cases.

In this section we shall explain a modification of the tree diagram
(see § 4 of Chapter Five). In this modification, we again represent each
state by a point, and each transition from one state to another by a directed
line or curve. The difference is that each state appears exactly once on
the graph, whereas in the tree diagram as we have been using it a given
state appears as often as it occurs in distinct sequences of pourings.

In order to illustrate the difference in these representations, we have
drawn them for the simple pouring problem with containers of capacities
3, 2, and 1 quarts, and three quarts of liquid initially in the largest container. Fig. 1 is the tree diagram, and Fig. 2 is the modified graph. The
states in Fig. 2 are arranged as they would appear on a (y, z) plane, in the
manner of Fig. 6 in Chapter Five, except that the vertices are labelled

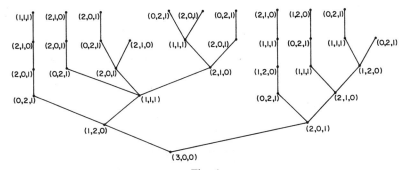

Fig. 1

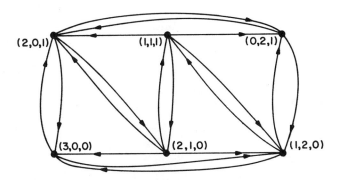

Fig. 2

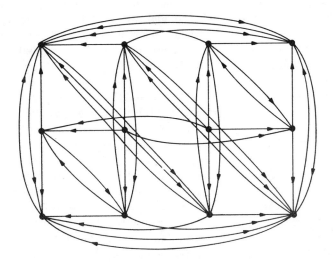

Fig. 3

with all three coordinates (x, y, z). As can be seen, Fig. 2 is more compact. On the other hand, Fig. 1 has the advantage of providing a visual display of all possible sequences of pouring operations, something which is not evident in Fig. 2. In more complex cases, the size of the tree diagram becomes a serious disadvantage, but on the other hand a graph of states and transitions like Fig. 2 becomes cluttered with many lines. For example, the corresponding graph for the pouring problem with jugs holding 5, 3, and 2 quarts is shown in Fig, 3. In Fig. 3, we have given up using the special symbol of Chapter Five, Diagram 1, to indicate that paths do not intersect, in order to simplify the appearance of the graph. Thus, all vertices are indicated by circles, and crossings of edges at other points are not to be considered vertices.

A simpler graph can be obtained if one merely suggests some of the edges, but does not draw them in full. Thus, Fig. 2 can be replaced by Fig. 4. In this figure, the six states are arranged in a more convenient pattern, four in an outside *ring*, and two down a central *spine*. There are two-way paths going all around the ring, and up and down the spine. In addition, there are one-way paths *leaving* the states along the spine and terminating at states on the outer ring. We use these conventions:

> ⟶connection to the state in the ring on the right
> ⟵connection to the state in the ring on the left
> ↑— or ⟶ connection to the state in the ring at the top
> ↓— or, ⟶ connection to the state in the ring at the bottom

Similarly, Fig. 3 can be replaced by Fig. 5, provided we omit the inaccessible states (3, 1, 1) and (2, 2, 1). As we know, the inaccessible states can never be reached from another state and are accordingly of lesser interest. On the other hand, it is possible to get from the inaccessible states to certain of the states shown in Fig. 5. The two states not shown could therefore be connected to the diagram, but by "one-way streets" only. We shall call graphs such as those in Figs. 4 or 5 Sawyer graphs, after their inventor.

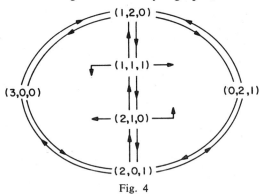

Fig. 4

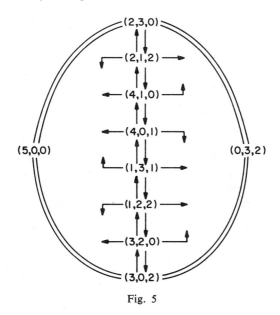

Fig. 5

2. Analysis of the Sawyer Graph

By careful analysis of the Sawyer graph, we can prove a number of general results concerning pouring puzzles. For example, we shall discuss the question of whether all the states on the boundary of the (y, z) diagram are completely accessible and we shall make some deductions concerning the best sequence of pourings to achieve a desired state.

In order to deduce *general* results, we must deal with containers of arbitrary sizes. Let us therefore consider the problem with three containers called A, B, and C which hold a, b, and c quarts, respectively. We do not specify any particular numerical values for a, b, c, but of course we do make certain obvious restrictions. One is that

$$a, b, c \text{ are positive integers} \tag{1}$$

In order to have a case of the type thus far dealt with and to avoid any additional complications, we also assume that the largest container holds as much as the two smaller ones combined. That is,

$$a = b + c \tag{2}$$

Moreover, we assume that one of the two smaller containers is bigger than the other and we may as well call it B. This is expressed algebraically by the inequality

$$b > c \qquad\qquad (3)$$

Finally, we assume that initially there are a quarts of liquid in the containers.

As usual, we shall denote a typical state of the system by the symbol (x, y, z) or (y, z), where x, y, and z are integers indicating the number of quarts of liquid in A, B, and C, respectively. The amount in A is $x = a - y - z$. Since a, b, and c do not have definite values, the (y, z) *state diagram* has to be drawn in a symbolic way such as in Fig. 6. In this figure, the dots between the number 2 and the expression $b - 1$ indicate that there are an unspecified number of integers between 2 and $b - 1$. The dots between the number 2 and the expression $c - 1$ have a similar significance. As usual, small dots or circles indicate possible states.

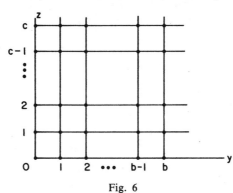

Fig. 6

We have already seen that interior points in Fig. 6 correspond to inaccessible states, since they represent states with no container either full or empty. Thus, we shall consider only the boundary or possibly accessible points. Furthermore, we shall divide these points into three categories as follows.

    I.   The states $(a, 0, 0)$ and $(0, b, c)$, in which every container is either full or empty. These states are represented by the points $(0, 0)$ and (b, c) in Fig. 6.

    II.   The states $(c, b, 0)$ and $(b, 0, c)$, in which two containers are either full or empty. (In each case, one is full and one is empty.) These states are represented by the points $(b, 0)$ and $(0, c)$ in Fig. 6.

    III.   The remaining states, in which just one container is either full or empty. These states are represented by the points in Fig. 6 which are on the outside boundary but not at the corners.

The Sawyer graph is now drawn in the following way. The states in class I are put at the left and right and those in class II at the top and bottom. Since the transitions

$$(a, 0, 0) \rightleftharpoons (c, b, 0)$$
$$(a, 0, 0) \rightleftharpoons (b, 0, c)$$

and

$$(c, b, 0) \rightleftharpoons (0, b, c)$$
$$(b, 0, c) \rightleftharpoons (0, b, c)$$

are easily seen to be possible, the four states in classes I and II from a *ring* like that in Fig. 5. We shall soon show that the states in class III can be arranged in a central *spine* like that in Fig. 5. Consequently, the Sawyer graph will have the general appearance of Fig. 7.

Our next step in the analysis of the general pouring problem is to show that from each state in class III exactly four transitions are possible, two reversible and two not reversible. To see this, we observe that to make a transition from any state we must choose two containers, one as donor and one as recipient. There are six ways to make this choice. But we cannot pour from the empty jug (if any) into either of the others, nor into the full jug (if any) from either of the others. Since one of these cases occurs, the number of outgoing links from any state in class III is exactly four. Moreover, a pouring operation is reversible if the donor is initially full, or the recipient initially empty, and not otherwise. Therefore there are two reversible operations for a given state of class III, from the initially full container (if any) into each of the others, or into the initially empty container (if any) from each of the others.

Now suppose that we begin with the state $(c, b, 0)$ at the top of the Sawyer graph. There are three possible operations, the results being $(a, 0, 0)$, $(0, b, c)$, and $(c, b - c, c)$. The first two are class I states on the ring

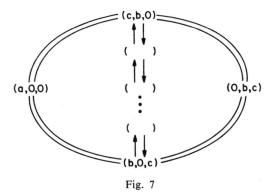

Fig. 7

and the last is a class III state on the spine of the graph. Since there are two reversible operations from each class III state, we see that a chain formed from reversible transitions between class III states cannot branch, nor can it end in any way except by reaching a state on the outer ring. Moreover a chain starting at $(c, b, 0)$ cannot terminate at $(a, 0, 0)$ or $(0, b, c)$ before reaching $(b, 0, c)$, since from each of $(a, 0, 0)$ and $(0, b, c)$ we have seen that there are exactly two reversible steps and these go to $(c, b, 0)$ and $(b, 0, c)$. We can therefore conclude that the chain of class III states which begins

$$(c, b, 0)$$
$$\downarrow \uparrow$$
$$(c, b - c, c)$$
$$\downarrow \uparrow$$

must connect to the state $(b, 0, c)$, thus generating the central spine of the Sawyer graph.

It does not follow, however, that the central spine contains *all* class III states. There might be one or more separate rings composed entirely of class III states and not connected to the others, as illustrated in Fig. 8.

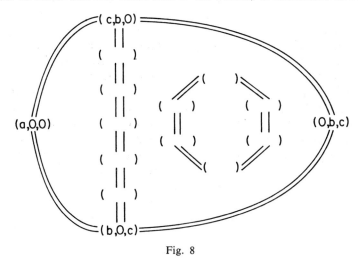

Fig. 8

Exercise

1. Construct the Sawyer graphs for the following jug sizes:

    (a)　7, 5, 2　　(b)　7, 4, 3
    (c)　9, 5, 4　　(d)　9, 6, 3

3.　Connectedness of the Sawyer Graph

We shall now prove that under certain conditions to be spelled out below the Sawyer graph is *connected,* rather than being in separate pieces as in Fig. 8. To do so, it is convenient to subdivide the states of class III. In the first two subclasses, one jug is empty and no jug is full:

(1)   states $(x, 0, a - x)$, where x is a number in the range $b < x < a$;

(2)   states $(x, a - x, 0)$, where x is a number in the range $c < x < a$.

There is no need to include a third subclass with the first jug empty since the only possible state of this type is in class I. The restriction $b < x$ in (1) is deduced from the fact that the third container is not full, which implies that $a - x < c$ or since $a - c = b$ that $x > b$. The restriction $c < x$ in (2) is deduced similarly. In the remaining two subclasses of class III, one jug is full and no jug is empty:

(3)   states $(x, b, c - x)$, where x is a number in the range $0 < x < c$;

(4)   states $(x, b - x, c)$, where x is a number in the range $0 < x < b$.

4.   The Spine of the Sawyer Graph

Now let us examine the sequence of pourings which begins from $(c, b, 0)$ at the top of the Sawyer graph. The first step is to $(c, b - c, c)$, which is a state of type III (4) since the last jug is full. From $(c, b - c, c)$, a reversible step must be a pouring from the last (full) jug into one of the others. One of the possibilities is to pour into the middle jug, leading back to $(c, b, 0)$. The other is to pour into the first jug, giving the state $(2c, b - c, 0)$. The first jug will not overflow in this operation since $2c < b + c = a$. Thus, the first two steps down the spine of the Sawyer graph are as follows:

State	Class
$(c, b, 0)$	II
$\updownarrow$	
$(c, b-c, c)$	III(4)
$\updownarrow$	
$(2c, b-c, 0)$	III(2)

Let us now see what subsequent operations will occur. From any state in class (2), say $(x, a - x, 0)$, the two possible reversible operations are to pour from the first or second jug into the last. Thus, from $(2c, b - c, 0)$ one can obtain $(c, b - c, c)$ by pouring from the first jug. This takes us back *up* the spine of the Sawyer graph. Hence, pouring from the middle jug is the reversible operation which yields a new state. In the present case, the result is the state $(2c, b - 2c, c)$, provided $b - 2c \geq 0$.

This new state is of type (4) again. From a type (4) state, say $(x, b - x, c)$, the two reversible operations are pourings from the last jug into one of the others. In our case, we have just poured from the middle jug into the last, and so pouring from the last into the first is the reversible operation which yields a new state.

From this discussion we conclude that the spine of the Sawyer graph will begin with a sequence of states alternately of types (4) and (2). The transitions between them correspond to pouring from the middle jug into the empty third jug and pouring from the full third jug into the first jug.

In Fig. 9 below, we indicate the general situation and include a record of the changes in content of the three jugs at each step. Thus each time a new pair of states is added to the chain, the content of the first jug goes up by c and the content of the middle jug goes down by c. Evidently these steps can be carried out as long as the content of the first jug does not exceed a and the amount withdrawn from the middle jug does not exceed b.

5.   The Case When c is a Divisor of b

If we let n be the largest integer such that $nc \leq a$, or equivalently $(n - 1)c \leq b$, then the steps shown in Fig. 9 are possible, but it is not possible to withdraw another c quarts from the middle jug. If a/c is an integer, or equivalently $(n - 1)c = b$, then the last two states shown are of classes II and I and are on the ring of the Sawyer graph. In this case, there are $(n - 2)$ pairs of states of types (4) and (2) and therfore there are $2n$ states in all on this connected part of the Sawyer graph. As shown in the Exercises below, there may be other detached chains of states.

State	Class	Change in		
		x	y	z
$(c, b, 0)$	II	—	—	—
$\downarrow\uparrow$				
$(c, b - c, c)$	III(4)	0	$-c$	$+c$
$\downarrow\uparrow$				
$(2c, b - c, 0)$	III(2)	$+c$	0	$-c$
$\cdots$	$\cdots$	$\cdots$		
$((n - 1)c, b - (n - 1)c, c)$	III(4) or II	0	$-c$	$+c$
$\downarrow\uparrow$				
$(nc, b - (n - 1)c, 0)$	III(2) or I	$+c$	0	$-c$

Fig. 9

ι. The General Case

If a/c is not an integer, then $nc < a < (n + 1)c$. The last two states in the above chain lie in classes (4) and (2). However, it is not possible to withdraw a full c quarts from the middle container. The next reversible step consists in emptying the middle container into the last one, yielding a state of type III(1), namely $(nc, 0, b - (n - 1)c)$. Now from a type (1) state, reversible steps are pourings into the middle container. The one of these yielding a new state is the pouring from the first container. Since $nc > a - c = b$, the middle jug can be filled and the new state is $(nc - b, b, b - (n - 1)c)$. This is of type (3). The next step, pouring from the full jug into the last jug, results in $(nc - b, 2b - nc, c)$, and the next step yields $((n + 1)c - b, 2b - nc, 0)$. Since $2b - nc = b - (nc - b) > b - c > 0$, and $(n + 1)c - b < a + c - b < a$, the last state exists and is of type (2). It is now clear that we shall again have a sequence of states alternately of types (4) and (2). The previous Fig. 9 can now be extended to Fig. 10.

Evidently this chain continues until the middle jug contains an amount y with $0 \leq y \leq c$. It will become empty if some multiple of c is exactly $2b$, say for example $(m - 1)c = 2b$. In this case, the chain ends with the states

$$((m - 1)c - b, 2b - (m - 1)c, c) \quad \text{or} \quad (b, 0, c)$$
$$(mc - b, 2b - (m - 1)c, 0) \quad \text{or} \quad (a, 0, 0)$$

The total number of pairs of states on the ring and spine is then $(m - 1) + 1 + (m - n + 1) = m + 1$ and the total number of states is $2(m + 1) = 4a/c$.

If no multiple of c is $2b$, we let m be the largest integer such that $(m - 1)c < 2b < mc$, or equivalently $0 < 2b - (m - 1)c < c$. Then from the state $(mc - b, 2b - (m - 1)c, 0)$ the next transition will be to a state of type (1) followed by a state of type (3), as indicated in Fig. 11.

It is now apparent that the spine of the Sawyer graph consists of chain of states such as

$$(4) \to (2) \to \cdots \to (4) \to (2) \to (1) \to (3) \to (4) \to (2) \to$$
$$\cdots \to (4) \to (2) \to (1) \to (3) \to \cdots .$$

The first $(4) \to (2)$ chain contains $n - 1$ pairs of states, the second contains $m - n$ pairs. Adding the pair of states of types (1) and (3), we find that the number of pairs of states in the spine down to the end of the second $(4) \to (2)$ chain is $n - 1 + 1 + m - n = m$. Likewise the next $(4) \to (2)$ chain will appear as follows:

$$(mc - 2b, 3b - mc, c) \qquad \text{III}(4)$$
$$\downarrow\uparrow$$

State	Class	x	y	z
			Change in	
$(c, b, 0)$	II	−	−	−
↓↑				
$(c, b − c, c)$	III(4)	0	−c	+c
↓↑				
$(2c, b − c, 0)$	III(2)	+c	0	−c
...	...		...	
$((n − 1)c, b − (n − 1)c, c)$	III(4)	0	−c	+c
↓↑				
$(nc, b − (n − 1)c, 0)$	III(2)	+c	0	−c
↓↑				
$(nc, 0, b − (n − 1)c)$	III(1)	0		
↓↑				
$(nc − b, b, b − (n − 1)c)$	III(3)	−b		
↓↑				
$(nc − b, 2b − nc, c)$	III(4)	0		
↓↑				
$((n + 1)c − b, 2b − nc, 0)$	III(2)	+c	0	−c
↓↑				
$((n + 1)c − b, 2b − (n + 1)c, c)$	III(4)	0	−c	+c
↓↑				
$((n + 2)c − b, 2b − (n + 1)c, 0)$	III(2)	+c	0	−c
...	...		...	

Fig. 10

State	Class	x	y	z
$((m − 1)c − b, 2b − (m − 1)c, c)$	III(4)	0	−c	+c
↓↑				
$(mc − b, 2b − (m − 1)c, 0)$	III(2)	+c	0	−c
↓↑				
$(mc − b, 0, 2b − (m − 1)c)$	III(1)	0		
↓↑				
$(mc − 2b, b, 2b − (m − 1)c)$	III(3)	−b		
↓↑				
$(mc − 2b, 3b − mc, c)$	III(4)	0		
↓↑				
$((m + 1)c − 2b, 3b − mc, 0)$	III(2)	+c	0	−c

Fig. 11

$$((m + 1)c - 2b, 3b - mc, 0) \qquad \text{III}(2)$$

$$\cdots$$

$$((k - 1)c - 2b, 3b - (k - 1)c, c) \qquad \text{III}(4)$$

$$\downarrow\uparrow$$

$$(kc - 2b, 3b - (k - 1)c, 0) \qquad \text{III}(2)$$

where k is the greatest integer such that $0 < 3b - (k - 1)c < c$, or $(k - 1)c < 3b < kc$. The total number of pairs of states down to this point will be $m + 1 + k - m = k + 1$. At the end of the next $(4) \rightarrow (2)$ chain the number of states will be $j + 2$ where j is the greatest integer such that $0 < 4b - (j - 1)c < c$, or $(j - 1)c < 4b < jc$. Suppose that N is the smallest nonnegative integer such that c divides $(N + 1)b$. Then we shall come to a final $(4) \rightarrow (2)$ chain, the final states being, let us say,

$$((p - 1)c - Nb, (N + 1)b - (p - 1)c, c) \qquad \text{II}$$

$$\downarrow\uparrow$$

$$(pc - Nb, (N + 1)b - (p - 1)c, 0) \qquad \text{I}$$

where $N \geq 0$ and $0 = (N + 1)b - (p - 1)c$, or

$$(p - 1)c = (N + 1)b \qquad (1)$$

The total number of pairs of states in the $(4) \rightarrow (2) \rightarrow \cdots \rightarrow (1) \rightarrow (3) \rightarrow \cdots$ sequences is now $p + N - 1$. However, the last two states are in fact $(b, 0, c)$ and $(a, 0, 0)$, which are on the ring. Adding one more pair of ring states we find this result:

Theorem. *Let N be the smallest nonnegative integer such that c divides $(N + 1)b$, and let p be defined by (1). Then the total number of states on the ring and spine of the Sawyer graph is $2(p + N) = 2(N + 1)a/c$.*

Let us illustrate the use of this result. Suppose that the three containers hold 8, 5, and 3 quarts, respectively, so that $a = 8, b = 5, c = 3$. We look for the smallest nonnegative integer N such that 3 divides $5(N + 1)$. This integer is $N = 2$, and the corresponding p is 6. The number of states is thus $2(p + N) = 16$, which is equal to $2a$, the number of possibly accessible states. Therefore the Sawyer graph is connected, and all states are either inaccessible or completely accessible.

7. The Case when b and c Have No Common Factor

If b and c have no common factor greater then one, the Sawyer graph is connected, and all states on the boundary of the (y, z) state diagram are completely accessible.

For in this case, (6.1) yields

$$p = 1 + (N + 1)b/c$$

Since b and c have no common factor, the smallest value of N yielding integral p is $N = c - 1$. Then $p = 1 + b$, $N - p = b + c = a$, and the number of states in the Sawyer graph is $2a$, which is the number of possibly accessible states.

If b and c have common factors, other then 1, the situation may be different, as indicated in the exercises.

Exercises

1. Use Eq. (6.1) to find N, p, and the number of states on the Sawyer ring and spine for each of the following jug sizes. Draw the complete Sawyer graph for (a) and (b).

    (a)   6,   4,   2   (b)   9,   6,   3
    (c)   16,   10,   6   (d)   27,   15,   12

2. Suppose that a is a multiple of c, $a = nc$. Show that $N = 0$ and $p = n$, so that the number of states on the ring and spine is $2(p + n) = 2n$. This agrees with the result in § 5.

3. Another way to find the number of states on the spine of the Sawyer graph is as follows. Note that each pair of states on the spine corresponds either to increasing the content of the large jug by c or decreasing it by b. Since $b + c = a$, a decrease of b is equivalent to an increase of c, *modulo a*. Therefore if there are $n - 1$ pairs in the spine, the total change is $(n - 1)c$, modulo a. On the other hand, going from $(c, b, 0)$ to $(b, 0, c)$ the change is $b - c$. Thus,

    $$(n - 1)c \equiv b - c \equiv -2c \pmod{a}$$
    $$(n + 1)c \equiv 0 \pmod{a}$$

    If b and c have no common factor, then a and c have no common factor. It then follows that

    $$n \equiv 1 \pmod{a}$$

    The smallest such integer n is $n = a - 1$. Thus, there are $a - 2$ pairs down the spine and $2a$ states in all on the Sawyer graph. This method can also be used to find the number of states on the spine if b and c have a common factor. (See any elementary number theory text for an explanation of congruences and modular arithmetic.)

4. Draw a Sawyer-like graph for a case in which $a > b + c$. Are the results of this section still applicable?

5. Assume that $a = b + c$, a is even, and that b and c have no common factor. Show that the state $(a/2, a/2, 0)$ can be reached from $(a, 0, 0)$ in $a - 1$ pourings, and cannot be reached in fewer pourings.

8. The Billiard Ball Solution

In this section we shall discuss the use of an analog computer to solve the pouring problem. First let us consider the pouring problem in which $a = b + c$ and $b \geq c \geq 0$. We construct the following diagram: Draw a parallelogram with a 60° angle in the lower left corner. Make the base b units long and the left edge c units long. Then the top will be b long and the right side c long. Mark off single units along the base and top and connect the marks to form lines parallel to the sides. Mark off each of the c units along the sides and connect them parallel to the base. Finally, draw lines at 120° angle to the base so that they connect the intersections made by other lines. Fig. 12 shows the diagram for the 8-5-3 pouring problem.

We label the intersections along the bottom and left edges as in Fig. 12 for clarity. In this diagram each intersection represents a different state. In fact, the diagram is essentially the same as Fig. 6 of Chapter Five, but by changing to this shape we can include more information in the diagram.

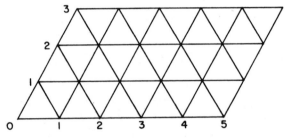

Fig. 12

The first thing we note about this new diagram is that it too has the inaccessible states on the inside and the possibly accessible ones along the edges. Next we note that the diagram has four types of intersection points: ones where two rays, three rays, four rays, and six rays join. The points where two rays join correspond to the states $(0, 0)$ and (b, c) each of which has two exits (there are two possible pourings from each). Three rays join at $(0, c)$ and $(b, 0)$ from which there are three exits. Similarly, where four rays join (boundary points other than vertices) the state has four exits, and where six rays join (interior points), the state has six exits. Although these six-exit states are all inaccessible, the diagram nevertheless contains information about them.

Starting at any state, we can find all the possible pourings from that state by following each line from that state to the line's terminus on the boundary. This terminal state will be the result of a pouring. We can prove this statement as follows: If we follow a horizontal line, we are leaving the amount in the c jug unchanged and are emptying or filling the b jug. If we travel along a $60°$ line, we are leaving the b jug unchanged and are emptying or filling the c jug. Finally, if we go along a $120°$ line, we are pouring from the b jug to the c jug or vice versa. Therefore, by following the lines we can go from any state only to a state which is accessible from the former state. To prove that we can get to all the states accessible from a given state, we remember that the number of lines emanating from a state equals the number of states accessible from that state. Therefore, we have a one-to-one correspondence between the set of moves along the lines and the set of pourings.

The case in which $a > b + c$ and $b \geq c > 0$ is set up the same way. The only difference is that for any state (y, z) there is more liquid in jug a than before. However, the case in which $a < b + c$ and $a > b \geq c > 0$ is a little different. For example, the state (b, c) is no longer possible. Fig. 13 shows the case 7-5-4. Again we find that all the accessible states are along the edges. We see that the line joining $(5, 2)$ and $(3, 4)$ contains all those states in which jug a is empty. These states are accessible. Furthermore, by arguing as before, we can prove that there is a one-to-one correspondence between the possible pourings from a state and the possible moves along the lines.

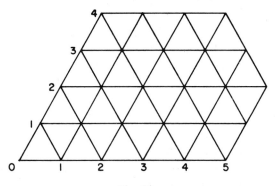

Fig. 13

9. The Billiard Ball Diagram and Sawyer Graph

By now some readers are probably wondering what relationship exists between the billiard ball diagram and the Sawyer graph. The answer is that the Sawyer graph can easily be constructed from the billiard ball diagram (for $a \geq b + c$; for $a < b + c$ we get a variation of the Sawyer graph). To generate the Sawyer graph, we proceed as follows. Start with a billiard ball in the $(0, 0)$ corner and shoot it along the base. When it hits $(b, 0)$ it will reflect along the ray going to $(b - c, c)$. Upon hitting at $(b - c, c)$, the ball reflects along the ray to $(b - c, 0)$. It is now apparent why we call this method the "billiard ball" technique and why we constructed the parallelogram with 60° and 120° corners. As the ball continues to bounce around, we record in order each of the points it strikes. These points form the main spine of the Sawyer graph. In the case where $a \geq b + c$ the ball will eventually come to the point $(0, c)$ from the point $(c, 0)$. Here it has a choice of two ways in which to "reflect." This tells us that the main spine of the Sawyer graph ends here. One reflection takes us back to $(0, 0)$ from which there are no more reflections. The other reflection takes us to (b, c) where the ball stops. Thus, we leave out only one two-way path between states, the one between $(b, 0)$ and (b, c). We can get this one by starting from $(0, 0)$ to $(0, c)$ instead of to $(b, 0)$. The one-way paths in the Sawyer graph are easily determined by inspecting the parallelogram. Each accessible point has two one-way exits to the vertices lying on its edge. For example, from $(b - c, c)$ there are one-way transitions to $(0, b)$ and (b, c).

The determination of a Sawyer-like graph for the case with $a < b + c$ is slightly more complicated. This is because there are now four corners having three exits. It is these corners which allow the choice of reflections. The best policy for constructing the Sawyer-like graph is as follows: Begin the process as before. When a corner offering a choice of reflections is first encountered, choose the path which goes to a point not already visited. Thereafter, when there is a choice of reflections, make the choice with the following order of preference: take a path leading to a new point: take a path along a new edge. To check this sequence, repeat the process starting to $(0, c)$ instead of $(b, 0)$. The one-way paths can be determined as before.

10.   Graphical Determination of Shortest Paths

To conclude our discussion of the "billiard ball" technique we shall present a graphical method for finding the shortest route from $(0, 0)$ to any accessible state. Suppose we wish to go from $(0, 0)$ to (y, z), where $0 \leq y \leq c$. If (y, z) is of the form $(b, 0)$ or $(0, c)$, the solution is apparent. If (y, z) is of the form $(b, a - b)$ or $(a - c, c)$, the solution is $(0, 0)$ $\rightarrow (b, 0) \rightarrow (b, a - b)$ or $(0, 0) \rightarrow (0, c) \rightarrow (a - c, c)$. All the remaining states (y, z) have four exits. Two of these exits are along an edge and two are through the interior to a different edge. Start from (y, z) and examine the two paths through the interior. If either ends on a corner state, the solution is to go to that corner as enumerated above and from there go to (y, z). If neither path leads to a corner, follow the path of reflection off the edge. This will be a path through the interior. If either ends at a corner, the problem is solved; if not, then keep repeating the process until a corner is reached or the paths both return to (y, z). In this latter case (y, z) is inaccessible from $(0, 0)$.

To make the process clearer, we shall give an example. Consider the problem with 9-, 7-, and 4-quart jugs. We first construct the billiard ball diagram as in Fig. 14. Suppose we wish to find the shortest route from $(0, 0)$ to $(0, 1)$. [This is equivalent to the fewest pourings to go from state $(0, 0)$ to $(0, 1)$.] We start by locating $(0, 1)$ and drawing in the two-way paths from $(0, 1)$. Since neither of these terminates at a vertex, we draw in the paths which are the reflection of the first paths. We continue until we reach a vertex (see Fig. 15). In this example the fourth ray of each path intersects a vertex. However, we know that $(0, 4)$ is one step (pouring) from $(0, 0)$, while $(5, 4)$ is two steps away. Therefore, the path of minimal length is $(0, 0) \to (0, 4) \to (4, 0) \to (4, 4) \to (7, 1) \to (0, 1)$.

This is a five-step path which took eight steps to determine (we drew

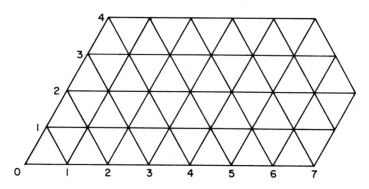

Fig. 14

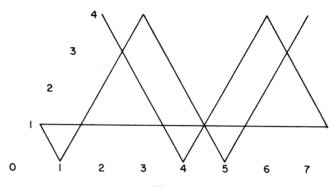

Fig. 15

eight lines). If we had drawn the ray from $(4, 0)$ to $(0, 4)$ before the one from $(5, 0)$ to $(5, 4)$, we would have saved one step, as we now show. We can demonstrate this as follows. Since neither of the third pair of rays drawn reached a vertex, there exists no four-step path. Thus, the best possible path is a five-step one. If we find a five-step path by drawing the first of our fourth pair of rays, we do not need to draw the other one. Similarly, if we wished to find the shortest path from $(0, 0)$ to $(1, 0)$, we would need to draw six lines to reach the vertex $(5, 4)$. We would then know the shortest path $(0, 0) \rightarrow (0, 4) \rightarrow (5, 4) \rightarrow (5, 0) \rightarrow (1, 4) \rightarrow (1, 0)$ and know its length, five steps. In this case it takes us six steps to find a five-step path. In general, if n is the minimum path length in steps, its determination takes $2(n - 1)$ or $2(n - 1) - 1$ steps when we stop our search on a vertex of form $(b, 0)$ or $(0, c)$, and $2(n - 2)$ steps when we finish our search on a vertex of form $(b, a - b)$ or $(a - c, c)$.* Thus, this search technique requires at least $2n - 4$ and at most $2n - 2$ steps.

The reader will see that the process described here for finding the shortest path to a given vertex is another version of the "labelling algorithm" described in Chapter Five. In Fig. 15, for example, drawing the first two lines from $(0, 1)$ gives us the pair of vertices $(1, 0)$ and $(7, 1)$ from which $(0, 1)$ can be reached in one step. The next two lines gives us $(1, 4)$ and $(4, 4)$, two steps from $(0, 1)$. Thus, the billiard ball method enables us to generate the sets S_1, S_2, and so on, as defined in § 14 of Chapter Five (the inaccessible states in S_1, S_2, etc., can be found too).

Exercises

1. Use the billiard ball technique to construct the Sawyer graphs for the following jug sizes:

   (a)   9, 5, 4   (b)   9, 6, 3   (c)   12, 7, 4
   (d)   10, 5, 3   (e)   12, 9, 7   (f)   11, 8, 5

2. Use the billiard ball technique to find the shortest path from $(0, 0)$ to:

   (a) $(0, 3)$ for jugs   11, 6, 5
   (b) $(6, 0)$ for jugs   13, 7, 5
   (c) $(14, 0)$ for jugs   23, 17, 11.

3. In § 3, states of class III were subdivided into types (1), (2), (3), (4). Where do these types appear on the billiard ball diagram? Explain on the basis of the billiard ball diagram why the sequence of types is $(4) \rightarrow (2) \cdots (4) \rightarrow (2) \rightarrow (1) \rightarrow (3) \rightarrow (4) \rightarrow (2) \rightarrow \cdots$.

4. Why can we never complete our search for a shortest route by the

---

* See Exercise 4.

method of § 10 in $2n - 5$ steps (where n is the length of the minimal path)?

5. Consider the problem with $a = b + c$ and $(b, c) = n > 1$ (n is the greatest common divisor of b and c). In this case many of the points of the boundary of the state diagram are inaccessible from $(0, 0)$. But these points do form a number of separate sets: each point of one of these sets is accessible only from other points of this set. How many such sets exist? Repeat the problem for $a > b + c$ and $a < b + c$, $(b, c) = n > 1$.

11. Discussion

Since we have mentioned that that billiard ball technique is an analog computer, it many be well to add a word concerning the difference between analog and digital computers. A modern digital computer operates internally by sending pulses of electricity to designated places at prescribed times. By using certain devices it is possible to count trains of pulses, and thus to perform addition and other arithmetic operations. The basic problem in using a digital computer is therefore to reduce the problem at hand to simple arithmetic (though ordinarily the amount of arithmetic to be done will be prodigious). On the other hand, an analog computer is a physical device whose operation imitates, or is parallel to that of another physical device. As we have shown that is what the billiard ball method does—the motion of the billiard ball simulates the pouring of liquid from jug to jug.

The slide rule is sometimes said to be an analog computer, because its motions are parallel to or in correspondence with certain arithmetic operations.

12. Computer Assistance in Drawing State Graphs

As we have seen, the Sawyer diagram and the billiard ball computer are very useful in discussing accessibility and connectedness, and in determining an optimal sequence of pourings to achieve a desired state. However, for extremely large puzzles (for example, $a = 269, b = 137, c = 132$) it may be impractical to attempt to construct these diagrams by hand. In such a case, it may be possible to obtain the desired information by one of the computational methods of Chapter Five (for example, see § 19).

Another interesting possibility is to use the computer as an aid in the actual construction of the diagrams. For example, looking back over Chapter Six we see that our analysis shows exactly how to form the sequence of states lying on the main spine of the Sawyer diagram. Therefore a computer can readily be programmed to generate and print-out the triples representing these states. With some additional effort, the print-out can even include double lines connecting successive states, as in our hand-drawn Sawyer diagrams.

We want to emphasize that the use of the digital computer in the construction of state graphs, or other types of descriptive charts or graphs, is of widespread importance and is by no means restricted to the particular instance given here. Other examples are given in the Exercises in the following chapters.

Exercises

1. Construct a flow chart and program to print out in a vertical column the states on the spine of the Sawyer diagram for the case $a = b + c$. Use this in drawing the Sawyer diagram for the 269–137–132 puzzle.

2. Write a program to compute the number of states on the ring and spine of the Sawyer graph (Refer to the Theorem in § 6).

3. Extend the program of Exercise 1 so that it will not only list the states on the ring and spine, but also all other cycles of states. Apply to Exercise 1 in § 7.

4. Discuss computational and graphical methods for pouring puzzles involving 4 containers (See the book of O'Beirne cited in the Bibliography).

Miscellaneous Exercises

Preliminary. There are a number of arithmetic puzzles which may be put in the following general form: "Given a sequence of positive integers $[a_1, a_2, \cdots, a_N]$ and a set of permissible operations upon these integers, such as addition, subtraction, multiplication, use of a factorial, taking of a square root, etc., it is required to obtain representations of all integers, n, in the form

$$n = R_1(a_1 R_2(a_2 \cdots R_N a_N) \cdots)$$

where $R_1, R_2 \cdots$, denote permissible operations.

We shall begin by considering the case where the only operations allowed are addition and multiplication.

1. Let $\{a_i\}$, $i = 1, 2, \cdots$, be a given set of positive integers and let $f_N(n)$ denote the number of ways in which a positive integer n can be represented in the form

$$n = a_1 * a_2 * a_3 * \cdots * a_N$$

where the a_i must be used in the designated order and where each asterisk is a $+$ or a $\times$.

Considering the case $n = 6$, show that multiple representations exist.

2. Starting from the end and considering the possibility of the last asterisk being the first $+$, the next to last being the first $+$, and so on, show that

$$f_N(n) = f_{N-1}(n - a_N) + f_{N-2}(n - a_N a_{N-1}) + \cdots$$
$$+ f_1(n - a_N a_{N-1} \cdots a_3 a_2) + g(n)$$

where

$$g(n) = 1 \quad \text{if} \quad n = a_1 a_2 \cdots a_N$$
$$= 0 \quad \text{if} \quad n \neq a_1 a_2 \cdots a_N$$

3. Obtain in this way the representation

$$1 + 2 + 3 + 4 + 5 + 6 + 7 + 8 \times 9 = 100$$

This representation was given in

W. Ley, *For Your Information*, Galaxy Magazine, Aug., 1961, 141.

4. Let $F_N(n) =$ minimum total number of $+$ signs in the foregoing representation of n. Let $F_N(n) = + \infty$ if no such representation exists. Show that

$$F_N(n) = \min [1 + f_{N-1}(n - a_N),$$
$$\min_{1 \leq r \leq N-2} F_{N-r-1}(n - a_N a_{N-1} \cdots a_{N-r})$$

5. Compute $F_9(100)$ in this way in the following steps

$$F_9(100) = \min [1 + F_8(91), \quad 1 + F_7(28)]$$
$$F_8(91) = \min [1 + F_7(83), \quad 1 + F_6(35)]$$
$$F_7(83) = \min [1 + F_6(76), \quad 1 + F_5(41)]$$
$$F_6(76) = \min [1 + F_5(70), \quad 1 + F_4(46)]$$

and so on.

Show that $F_9(100) = 4$ and that $100 = 1 \cdot 2 \cdot 3 \cdot 4 + 5 + 6 + 7 \cdot 8 + 9$.

6. Are there any other representations of 100 apart from the two given above?

For the first six exercises, see

R. Bellman, "On Some Mathematical Recreations," *Amer. Math. Monthly* **69** (1962) 640–643.

7. If we allow $+$ and $-$ signs and proximity we can obtain many more representations. For example,

$$1 + 2 + 3 - 4 + 5 + 6 + 78 + 9 = 100$$
$$1 + 23 - 4 + 56 + 7 + 8 + 9 \quad = 100$$
$$12 + 3 + 4 + 5 - 6 - 7 + 89 \quad = 100$$
$$123 - 45 - 67 + 89 \qquad\qquad = 100$$

Are there any other representations?

8. Assume that we are given four 4's and that the ordinary arithmetic operations are permissible. What integers can we represent?

(Example) $1 = 4/4 + 4 - 4$, $\quad 2 = 4/4 + 4/4$,

$3 = (4 + 4 + 4)/4$, $\quad 4 = 4 + (4 - 4)$, etc.

Bibliography and Comment

The Sawyer graph was first described in

W. W. Sawyer, "On a Well-Known Puzzle," *Scripta Mathematica* **16** (1950) 107–110.

The solution by means of the billiard ball method was first published (but not so named) in

M. C. K. Tweedie, "A Graphical Method of Solving Tartaglian Measuring Puzzles," *Math. Gazette* **23** (1939), 278–282.

The "billiard ball" interpretation is due to Y. I. Perelman, according to N. Court. See his book, "Mathematics in Fun and in Earnest," previously cited.

Tweedie pointed out that the coordinates (x, y, z) describing the state of the jugs can be taken to be the "trilinear coordinates" of a point with respect to an equilateral triangle. See also

D. Pedoe, *The Gentle Art of Mathematics*, English Universities Press, 1958.

A good historical survey of pouring puzzles is given in

T. H. O'Beirne, *Puzzles and Paradoxes*, Oxford University Press, New York and London, 1965.

Chapter Seven

CANNIBALS AND MISSIONARIES

1. Another Classical Puzzle

In this chapter, we wish to consider the following classical conundrum: "A group consisting of three cannibals and three edible missionaries seeks to cross a river. A boat is available which will hold at most two people, and which can be navigated by any combination of cannibals and missionaries involving one or two people. If the missionaries on either side of the river, or in the boat, are outnumbered at any time by cannibals, dire consequences which may be guessed at will result. What schedule of crossings can be devised to permit the entire group of cannibals and missionaries to cross the river safely?"

Our experience with the wine-pouring puzzle suggests the desirability of looking at this puzzle, too, as a succession of transitions from one state, or condition, to another, and indeed such an interpretation is possible. In fact, in attempting to solve the puzzle we can imagine a sequence of operations such as the following:

(1)   send two cannibals across the river;
(2)   return one cannibal;
(3)   send two cannibals across the river;
(4)   return one cannibal;
(5)   send two missionaries across the river;

etc. Thus, each crossing of the river by boat can be regarded as an operation which changes the condition of the system. The condition, or state, of the system can once again be respresented by a set of numbers. If we let

m = number of missionaries on the first bank
c = number of cannibals on the first bank

then the pair (c, m) suffices to describe the situation at any time that the boat is not in midstream. It is not also necessary to give the number of missionaries and cannibals on the second bank of the river, since the total number of each must always be three (assuming that the number of missionaries does not suffer an unhappy diminution—a conservation relation again) Thus, (3, 3) indicates that there are three missionaries and three cannibals on the left bank and none on the right; whereas (1, 3) indicates that there are one cannibal and three missionaries on the left bank and two cannibals on the right bank. The five operations listed above can be shown symbolically as follows:

(1)   (3, 3) $\longrightarrow$ (1, 3)
(2)   (1, 3) $\longrightarrow$ (2, 3)
(3)   (2, 3) $\longrightarrow$ (0, 3)
(4)   (0, 3) $\longrightarrow$ (1, 3)
(5)   (1, 3) $\longrightarrow$ (1, 1) .

This method of representing the state of the system by two numbers rather than four* is of course very helpful both from the standpoint of computer programs and of geometrical representations.

2. Trees and Enumeration

Most of the methods which we used to discuss the pouring puzzle have counterparts for the cannibal puzzle. For example, the method of enumeration by constructing a tree diagram can be carried over very easily. One must again be careful to use only allowable states. Thus, from (3, 3) the transition to (1, 3) is possible (by sending two cannibals across in the boat), but the transition from (3, 3) to (3, 1) is not allowable since it would

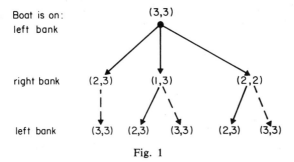

Fig. 1

* Or six if the occupants of the boat are also listed as part of the state.

result in disaster for the one missionary remaining on the left bank. In Fig. 1 we show the first part of a tree diagram, with all possible transitions shown. Those shown by dotted lines would normally be omitted since they lead back to a state already encountered. We leave it to the reader to complete the tree diagram, since we wish to go on to other methods.

3.   The State Diagram

A geometrical representation of the possible states, like that in §7 of Chapter Five is also possible. We need only plot every allowable pair (c, m) on a rectangular coordinate system. For the problem with three cannibals and three missionaries, c and m can be 0, 1, 2, or 3. There are therefore 16 possible pairs (c, m), but a number of these are not allowable. In fact, if $c > m > 0$ the state is not allowable, since there are more cannibals than missionaries on the first bank. On the other hand if $c > m$ but $m = 0$, the state *is* allowable, since there are no missionaries to be eaten. If $3 > m > c$, the state is also not allowable, since there are more cannibals than missionaries on the second bank. Thus, the set of allowable states consists of all pairs (c, m) satisfying these restrictions:

(a) $0 \leq c \leq 3$,
(b) $0 \leq m \leq 3$,
(c) $c = m$ or $m = 0$ or $m = 3$.

This set of states is depicted in Fig. 2. There are ten allowable states, lying in the shape of a large letter Z. As before, we shall call such a representation the *state diagram*.

Transitions between states can also be shown geometrically on the state diagram. For example, Fig. 3 shows the five ferryings listed in §1. The puzzle of the missionaries and cannibals can now be interpreted as calling for a sequence of transitions starting at the upper right corner of the state diagram and terminating at the lower left corner.

How will possible transitions appear on the state diagram? As we see from Fig. 3, they sometimes pass from one point to an adjacent point and

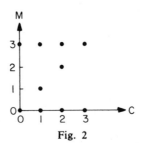

Fig. 2

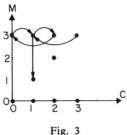

Fig. 3

sometimes jump over to a nonadjacent point (as they did in the pouring puzzle). In any case, passage of the boat from the first bank to the second bank reduces the values of c and m since it removes people from the first bank. The total decrease in c and m is either one or two since the capacity of the boat is two. Thus the only points which are reachable from a given point on such a passage lie in a right triangle with the given point at the upper right vertex, as in Fig. 4. Of course, although these points may be reachable, they may not all be allowable states. For example, from the starting point $(3, 3)$ in Fig. 2, it is possible and allowable to go to $(1, 3)$, $(2, 3)$, and $(2, 2)$, but not to $(3, 2)$ and $(3, 1)$, which are reachable but not allowable.

Similarly, passage of the boat from the second bank to the first bank corresponds to a motion upward and to the right on the state diagram. The points reachable from a given point in such a transition are shown in Fig. 5.

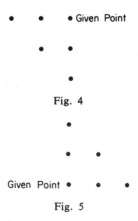

Fig. 4

Fig. 5

Keeping these rules in mind, we can construct a graph showing all states and all possible transitions. The result, shown in Fig. 6, will be called the *graph* of the puzzle. Each point representing a state will also be called a *vertex* of the graph, and each line joining two vertices will be called an *edge* of the graph. Each edge can be traversed in each direction, correspond-ing to the fact that all ferryings are reversible. It should be kept in mind that passages from the first to the second bank must alternate with passages from the second to the first bank. This means that on the graph edges leading down or to the left must alternate with edges leading up or to the right. Thus, (3, 3) to (2, 2) to (1, 1) to (0, 0) is not a permissible solution of the puzzle.

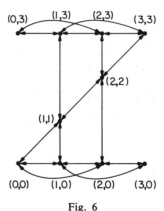

Fig. 6

A method for finding a solution to the puzzle can now be described as follows. Starting at the upper right corner, make a sequence of transitions to points in the reachable triangles, moving alternately down or to the left and up or to the right, until the lower left corner of the graph is reached. After a little experimentation, the solution shown in Fig. 7 is obtained. This solution requires 11 ferryings.

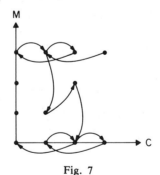

Fig. 7

4. Existence and Uniqueness

It is natural to ask whether the solution shown in Fig. 7 is the only solution to the puzzle. More generally, if one starts with some other number of cannibals and missionaries, is the puzzle solvable, and if so, which solution (if there are several) requires the fewest crossings of the river?

The graphical method is a help in answering these questions. Let us classify the allowable states into three classes:

(T) those along the top ($m = 3$)
(B) those along the bottom ($m = 0$)
(D) those along the diagonal ($0 < m = c < 3$)

Since the boat holds only two people, it is impossible to go directly from a state of type T to one of type B. Hence, any solution of the puzzle must include a type D state. Now if we move from a T state to the state $(2, 2)$, then at the next ferrying we must return to a T state. Therefore, in order to leave T states permanently, a transition must be made further down the diagonal than $(2, 2)$, namely to the state $(1, 1)$. The only way is from $(1, 3)$. Furthermore, the only way to reach B states is to go to $(2, 2)$ and $(2, 0)$. Thus, we have shown that a solution to the puzzle must contain the transitions shown:

$$\cdots \rightarrow (1, 3) \rightarrow (1, 1) \rightarrow (2, 2) \rightarrow (2, 0) \rightarrow \cdots .$$

Finally, it is easy to see that $(1, 3)$ can be reached through T states by

$$(3, 3) \rightarrow (1, 3) \rightarrow (2, 3) \rightarrow (0, 3) \rightarrow (1, 3)$$

or

$$(3, 3) \rightarrow (2, 2) \rightarrow (2, 3) \rightarrow (0, 3) \rightarrow (1, 3)$$

and similarly $(0, 0)$ can be reached from $(2, 0)$ in two ways. Hence, there are exactly four distinct solutions of the puzzle. However, the solution shown by Fig. 7 cannot be replaced by one with fewer steps.

Four cannibals and four missionaries cannot be taken safely across a river with a boat holding only two people. To see this, we refer to the state diagram in Fig. 8. In order to leave the T states without returning,

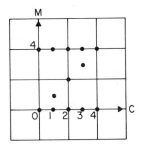

Fig. 8

one must go from $(2, 4)$ to $(2, 2)$. The next step must be to $(3, 3)$. Thereafter one cannot reach a *B* state, but can at best return to $(2, 2)$. There is no way to reach the bottom states, and indeed at most four of the eight persons can be transported across the river.

Let us consider the puzzle if the boat will hold three people. Remembering that the boat must not contain more cannibals than missionaries, we find that the reachable points are shown in Fig. 9. It is now possible to transport as many as five cannibals and five missionaries across the river. Fig. 10 shows how this may be done in 11 crossings. We leave it to the reader to show that it cannot be done in fewer crossings and also that it is not possible to transport six cannibals and six missionaries.

On the other hand, if the boat will hold four or more people, then any number of missionaries and cannibals can be transported across the river, for it is then possible to move down along the diagonal of the state diagram (as indicated in Fig. 11 for the case of six cannibals, six missionaries and a boat holding four). On the other hand, in some cases there is a solution using fewer crossings than that using just the diagonal states. For example, if there are six missionaries and cannibals and the boat holds five, the solution using the diagonal is still the one in Fig. 11, but there is a different solution requiring only seven crossings (see Exercise 3).

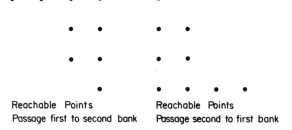

Reachable Points   Reachable Points
Passage first to second bank   Passage second to first bank

Fig. 9

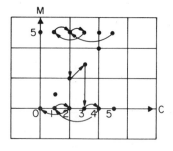

Fig. 10

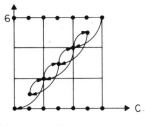

Fig. 11

By closer examination of the state diagrams, one can deduce various other results on the minimal number of crossings necessary as well as on a number of variants of the original puzzle. Several of these results are given in the Exercises below. We shall turn aside from these questions, however, in order to discuss analytical and computer techniques which could be used, and would be required, for more complex problems.

Exercises

1. Show that for the puzzle with n cannibals and n missionaries $(n > 3)$ and a boat holding two people, it is possible to transport all the cannibals across the river, but at most two missionaries.

2. Let M and C denote the number of missionaries and cannibals, respectively, originally on the first bank of the river, and let B denote the capacity of the boat. Draw the state diagram for $M = 5$, $C = 4$, $B = 2$ and show that the puzzle has a solution.

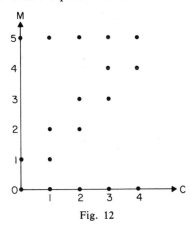

Fig. 12

3. Show that the puzzle with $M = C = 6$, $B = 5$ can be solved with only seven crossings.

4. Suppose that $M = 4$, $C = 2$, $B = 2$, but that cannibals are stronger than missionaries to the extent that the missionaries are safe only when they exceed the cannibals in number (on each bank and in the boat). Draw the state diagram and show that the party cannot cross safely.

5. Show that if $M = C$, B is even, and $4 \le B \le M$, the minimal number of crossings is obtained by following the diagonal of the state diagram.

6. Is the result in Exercise 5 correct if B is odd?

5. Analytic Methods

Although the graphical method is easy to use for the crossing puzzles so far considered, it may not be easy for more complex puzzles. In particular, this may be the case if more than two variables are required to denote the state of the system. In such cases, use of one of the other techniques presented below may be more systematic and quicker.

We shall explain one of the possible techniques as it could be applied to the present class of puzzles. The method we shall discuss is similar in spirit to that of § 18 of Chapter Five, in that we shall ask how close one can come to solving the cannibals and missionaries puzzle in those cases where a true solution is impossible. As usual, it is necessary to say just what "closeness" means here. Suppose that for the present we continue to look at the puzzle in which we can denote the state of the system by the ordered pair (c, m), where c is the number of cannibals and m is the number of missionaries on the first bank of the river. We let C and M respectively denote the total numbers of cannibals and missionaries. Then the "distance" between two states (c_1, m_1) and (c_2, m_2) could be defined as $|c_1 - c_2| + |m_1 - m_2|$. For example, the distance of (c_1, m_1) from the desired state $(0, 0)$ is $c_1 + m_1$, that is, it is the number of cannibals and missionaries remaining to be transported across the river.

Following the procedure of § 18, Chapter Five, we can now define the function*

$f_N(c, m) =$ the minimum number of people remaining on the first bank after exactly N allowable crossings, starting with c cannibals and m missionaries on the first bank and the boat on the first bank   (1)

For certain states (c, m), no transition to an allowable state is possible. Then $f_1(c, m)$, $f_2(c, m)$, etc., are undefined. However, if one crossing is

_____

* Alternatively $f_N(c, m)$ could be defined as the maximum number of people on the second bank after N crossings.

possible, so that $f_1(c, m)$ is defined, then, since every crossing is reversible, any number of crossings can occur and $f_2(c, m), f_3(c, m)$, etc., are all defined.

Also, let

$g_N(c, m) =$ the minimum number of people remaining on the first bank after exactly N allowable crossings, starting with c cannibals and m missionaries on the first bank and the boat on the second bank $\qquad (2)$

Again, $g_N(c, m)$ may be undefined for certain (c, m) for $N \geq 1$. Symbolically, these definitions can be written

$$f_N(c, m) = \min_{(c_1, m_1) \in Q_1} (c_1 + m_1) \qquad (3)$$

where Q_1 is the set of allowable states which can be reached from state (c, m) in exactly N crossings if the boat is initially on the first bank;

$$g_N(c, m) = \min_{(c_1, m_1) \in Q_2} (c_1 + m_1) \qquad (4)$$

where Q_2 is the set of states which can be reached from state (c, m) in N crossings if the boat is initially on the second bank. Q_1 is empty if and only if $f_N(c, m)$ is undefined and Q_2 is empty if and only if $g_N(c, m)$ is undefined.

Now let us start from (c, m) with the boat on the first bank. To achieve the minimum $f_N(c, m)$, we must send a certain number c_1 of cannibals and m_1 of missionaries across in the boat, and then continue with an optimal choice of $N - 1$ crossings starting from the resulting state $(c - c_1, m - m_1)$. Thus, using a familiar argument we obtain the relation

$$f_N(c, m) = \min_{c_1, m_1} g_{N-1}(c - c_1, m - m_1), \qquad N \geq 1 \qquad (5)$$

where the minimization is achieved by considering all choices of c_1 and m_1 which fit the various constraints. Similarly,

$$g_N(c, m) = \min_{c_1, m_1} f_{N-1}(c + c_1, m + m_1), \qquad N \geq 1 \qquad (6)$$

where c_1, m_1 now denote numbers sent back to the first bank. We also have

$$f_0(c, m) = c + m, \qquad g_0(c, m) = c + m \qquad (7)$$

If in (5) or (6) there are no choices of c_1 and m_1 which fit the constraints, then no improvement is possible, and $f_N(c, m)$, or $g_N(c, m)$, is undefined.

6. Example

Let us illustrate the use of these relations in a simple case, for example that in which $M = C = 3$ and the boat capacity is $B = 2$. Then, in using (5), one must have $0 \leq c_1 \leq c$, $0 \leq m_1 \leq m$ and $1 \leq c_1 + m_1 \leq 2$. In (6) one must have $0 \leq c_1 \leq 3 - c$, $0 \leq m_1 \leq 3 - m$, $1 \leq c_1 + m_1 \leq 2$. The choice of c_1 and m_1 is also constrained by requiring that the new state be allowable.

As in our work on the pouring puzzle, we find it helpful to construct a "cost matrix." Following a familiar procedure we list the ten possible states along the top and along the left edge of an array. In the ith row and jth column we place a 1 if it is possible to go from state j to state i in one crossing. But now there is a new feature: such a transition may be possible if the boat is on the first bank, but impossible if it is on the second bank, or vice versa. We can overcome this ambiguity in the following way. Let us form the cost matrix A based on the assumption that the boat is on the first bank. That is, the element a_{ij} in the ith row and jth column is defined by

$$a_{ij} = 1 \text{ if the transition from } j \text{ to } i \text{ is possible in one ferry-}$$
$$\text{ing when the boat starts on the first bank } (i \neq j) \qquad (1)$$

All other a_{ij} are left blank in the cost matrix (or set equal to zero in the connection matrix). But now we notice also that a_{ij} is one if and only if the transition in the reverse direction, from i to j, is possible in one ferrying, when the boat starts on the second bank. This is because a given ferrying from j to i can always be cancelled out by carrying the occupants of the boat back from i to j. Thus, the single matrix defined by (1) shows the possible transitions from first to second bank if we read it column to row, and from second to first bank if we read it row to column.* This "modified cost matrix" for the puzzle $M = C = 3$, $B = 2$, is shown in Table 1.

The calculation can now be set down in a paper strip algorithm as follows. We begin by computing $f_0(c, m)$ and $g_0(c, m)$ for all states (c, m), using Eq. (5.7). Now put the numbers $f_0(c, m)$ on a horizontal strip and the numbers $g_0(c, m)$ on a vertical strip. Hold the vertical strip next to a column of the cost matrix. Wherever there is an entry one in the matrix, a transition from the column state to the row state is possible by crossing from the first to the second bank. As we see from Eq. (5.5), the minimum of the numbers on the g_0 strip opposite these one's is then the value of f_1 for the state corresponding to the given column. The state in the row which yields the minimum is the minimizing state (c_1, m_1). If the column in the cost matrix is entirely blank, no transition to an allowable state is

* There is another way to say this: the matrix A^T, the transpose of A, shows the possible transitions from the second to the first bank if we read it column to row.

Example **207**

possible. In this event,' $f_1(c, m)$ is undefined. In this way, the values of $f_1(c, m)$ for all (c, m) are computed, and we record them as in Table 2. Next, the horizontal strip is placed next to a row in the cost matrix. The minimum number f_0 on the strip is sought among those next to a 1 in the cost matrix, for this means that it is possible to go from the row state to the column state in one ferrying from the second to the first bank of the river. The procedure therefore yields $g_1(c, m) = \min f_0(c + c_1, m + m_1)$. The resulting values $g_1(c, m)$ are recorded, and the procedure is now repeated by writing the numbers $f_1(c, m)$ and $g_1(c, m)$ on horizontal and vertical strips, respectively. Since we are constantly dealing with g_N on vertical strips and f_N on horizontal strips, the format in Table 2 is convenient. The reader should check and extend the results given in this table to make sure that he understands the procedure being used.

TABLE 1

MODIFIED COST MATRIX, $M=C=3$; $B=2$

	(0, 3)	(1, 3)	(2, 3)	(3, 3)	(2, 2)	(1, 1)	(0, 0)	(1, 0)	(2, 0)	(3, 0)
(0, 3)	1	1								
(1, 3)		1	1							
(2, 3)			1							
(3, 3)										
(2, 2)		1	1							
(1, 1)	1			1						
(0, 0)						1		1	1	
(1, 0)						1			1	1
(2, 0)			1							1
(3, 0)										

As we see from Table 2, the method given here leads to rather slow convergence. In fact, $f_k(c, m)$ becomes zero only after a number of steps equal to the number of crossings required to transport all $c + m$ people across the river. We can speed up the convergence by the same kinds of methods used before. For example, calculation of the g's can be eliminated by observing that

$$f_N(c, m) = \min_{c_i, m_i} f_{N-2}(c - c_1 + c_2, m - m_1 + m_2) \qquad (2)$$

where we now think of sending c_1 cannibals and m_1 missionaries from the first to the second bank and c_2 cannibals and m_2 missionaries back to the first bank. Calculation based on (2) is easier if a new array is constructed to show the states which can be reached from each given state in such a round trip.

TABLE 2

CALCULATION OF $f_N(c, m)$, $g_N(c, m)$; $C=M=3$, $B=2$

State (c, m)	$g_0(c, m)$	min state	$g_1(c, m)$	min state	$g_2(c, m)$	min state	$g_3(c, m)$	min state	$g_4(c, m)$
(0, 3)	3	(1, 3)	4	(1, 3)	2	(1, 3)	4	(1, 3)	2
(1, 3)	4	(2, 3)	5	(2, 3)	3	(2, 3)	4	(2, 3)	2
(2, 3)	5	(3, 3)	6	(3, 3)	4	(3, 3)	5	(3, 3)	3
(3, 3)	6	—	—	—	—	—	—	—	—
(2, 2)	4	(2, 3)	5	(2, 3)	3	(2, 3)	4	(2, 3)	2
(1, 1)	2	(1, 3)	4	(2, 2)	2	(2, 2)	3	(2, 2)	1
(0, 0)	0	(1, 0)	1	(1, 1)	0	(1, 1)	1	(1, 1)	0
(1, 0)	1	(1, 1)	2	(1, 1)	0	(1, 1)	1	(1, 1)	0
(2, 0)	2	(3, 0)	3	(3, 0)	1	(3, 0)	2	(3, 0)	0
(3, 0)	3	—	—	—	—	—	—	—	—

State (c, m)	(0, 3)	(1, 3)	(2, 3)	(3, 3)	(2, 2)	(1, 1)	(0, 0)	(1, 0)	(2, 0)	(3, 0)
$f_0(c, m)$	3	4	5	6	4	2	0	1	2	3
min state	—	(1, 1)	(0, 3)	(1, 3)	(1, 1)	(0, 0)	—	(0, 0)	(0, 0)	(1, 0)
$f_1(c, m)$	—	2	3	4	2	0	—	0	0	1
min state	—	(0, 3)	(0, 3)	(1, 3)	(2, 0)	(0, 0)	—	(0, 0)	(0, 0)	(1, 0)
$f_2(c, m)$	—	4	4	5	3	1	—	1	1	2
min state	—	(0, 3)	(0, 3)	(1, 3)	(2, 0)	(0, 0)	—	(0, 0)	(0, 0)	(1, 0)
$f_3(c, m)$	—	2	2	3	1	0	—	0	0	0
min state	—	(1, 1)	(0, 3)	(1, 3)	(2, 0)	(0, 0)	—	(0, 0)	(0, 0)	(1, 0)
$f_4(c, m)$	—	3	4	4	2	1	—	1	1	1

Exercises

1. Complete Table 2. In what sense does the process converge?

2. Construct an array which shows the possible transitions between states in one round trip, starting with the boat on the first bank. Using this array and Eq. (6.2), compute f_2, f_4, f_6, and so on.

3. Carry out the computational solution of the puzzle when $M = C = 4$ and $B = 2$.

4. Write a computer program for the calculation of $f_N(c, m)$ and $g_N(c, m)$ for all possible states (c, m), for each choice of C and M in the range $1 \leq C = M \leq 20$, when the capacity of the boat is $B = 5$. Compute for $N = 1$, 2, and so on until there is convergence or $N = 20$, say. Compare results with those obtained by the graphical method.

5.. Write a complete program to list all allowable states for the puzzle with arbitrary M, C and B subject to the restrictions $M \leq 20$, $C \leq 20$, $B < M$.

Miscellaneous Exercises

1. Consider the following scheduling problem, sometimes called the Hitchcock–Koopman–Kantorovich transportation problem.

Sources		Demands	
x_1	1	d_1	1
x_2	2	d_2	2
		$\vdots$	
		d_N	N

   At the two sources, 1 and 2, there exist supplies x_1 and x_2 of a particular item; at the demand points 1, 2, $\cdots$, N, there are demands d_1, d_2, $\cdots$, d_N respectively for these items. Let t_{ij} denote the distance from the ith source to the jth demand, and let the cost of shipping a quantity y_i from i to j be given by $y_i t_{ij}$. Assuming that the total supply equals the total demand, $x_1 + x_2 = d_1 + d_2 + \cdots + d_N$, determine the supply schedule which minimizes the total cost.
(Hint: Let $f_N(x_1, x_2)$ denote the minimum total cost of supply when there are N demand points, in a fixed order, and convert the scheduling problem into a multistage problem by assuming that we supply the demands of point N first, then point $N - 1$ and so on. Show that

$$f_N(x_1, x_2) = \min_{\{y_1 + y_2 = d_N\}} [t_{1N}y_1 + t_{2N}y_2 + f_{N-1}(x_1 - y_1, x_2 - y_2)]$$

for $N \geq 2$)

2. Examine the computational feasibility of the foregoing method.

3. Show that we can use the fact that total demand equals total supply to write $f_N(x_1, x_2) \equiv g_N(x_1)$.

4. Examine the computational feasibility in the light of this reduction. What advantages are there to using a sequence of functions of one variable as opposed to functions of two variables?

5. Consider the case where there is a "fixed cost," c_{ij}, of shipping any quantity y_i from i to j, with the total cost given by $c_{ij} + y_i t_{ij}$ if $y_i > 0$, and given by 0 if $y_1 = 0$. Does this change optimal scheduling policies? In what way?

6.   Considering the capacity of current computers, rapid access and disc storage, how many sources can reasonably be allowed?

7.   Develop a method of successive approximations for the case where there are 20 or more sources. In particular, can you devise an approximation scheme which guarantees to do at least as well as a scheduling policy determined by someone else?

For the foregoing, see

R. Bellman and S. Dreyfus, *Applied Dynamic Programming*, Princeton University Press, 1962.

8.   How shall a regiment cross a river when the only means of transportation is a boat containing two small boys? Either boy can operate the boat, but the boat is so small that it can carry at most one soldier or the two boys.

M. Kraitchik, *Mathematical Recreations*, Dover Publications, Inc., New York, 1953, 215.

9.   Two jealous husbands and their wives must cross a river in a boat that holds only two persons. How can this be done so that a wife is never left with the other woman's husband unless her own husband is present?

Ibid.

10. Consider the general case where the number of couples is $n \geq 2$ and the capacity of the boat is m, less than the total number of people. Show that

   a. $m = 4$ will suffice for $n \geq 3$
   b. $m = n - 1$ will suffice for $n \geq 3$
   c. $m = 2$ will not suffice for $n \geq 4$
   d. $m = 3$ will not suffice for $n \geq 6$
   e. $m = 3$ will suffice for $n = 5$

M. Kraitchik, *Mathematical Recreations*, Dover Publications, Inc., New York, 1942, Ch. 9.

11. Show that $m = 2$ will suffice for any number of couples if there is an island in the middle of the river.

Ibid.

12. Suppose that wives alone are unable to manage the boat?

Ibid.

13. Suppose that one man has two wives? What is the minimum number of trips required to transport four families?

Ibid.

14. Suppose that there are three cannibals, three missionaries and that the boat holds three people. Also suppose that the speed at which the boat can be rowed across the river is directly proportional to the number of people in the boat. What is the minimum *time* in which the entire party can be safely transported across the river?

15. What is the maximum distance into the desert which can be reached from a frontier settlement by the aid of a party of n explorers, each capable of carrying provisions that would last one man for a days?

16. If depots can be established, the longest possible journey will occupy
$$\frac{a}{2}\left(1 + \frac{1}{2} + \frac{1}{3} + \cdots + \frac{1}{n}\right) \text{ days.}$$

W. W. Rouse Ball, *Mathematical Recreations and Essays*, Macmillan, New York, 1947. 32.

17. Consider the game of Nim. Any number of chips are divided arbitrarily into a number of piles. Two people play alternately. Each, upon his turn, may select any one heap and remove from it all the counters in it, or as many of them as he pleases, but at least one. The player drawing the last counters wins. How does one determine optimal play?

Ibid., 36, 37.

18. (Moore's Nim) Consider the same game with the difference that each player can choose from at most k heaps on any move.

Ibid., 38.

19. (Wythoff's game) Consider the case where there are only two heaps and a player choosing from both heaps must take the same number from each heap. He can, however, choose counters from only one heap.

20. Consider the game of Nim where a player can only draw amounts $a_1, a_2, \cdots, a_k$ from a heap. See also

J. B. Haggerty, "Kalah—An Ancient Game of Mathematical Skill," *Arithmetic Teacher* (May, 1964) 326–330.

Time Magazine **81** (June 14, 1963) No. 24, 67.

C. A. B. Smith, "Graphs and Composite Games," *J. Combinatorial Theory* **1** (1966) 51–81.

Bibliography and Comment

For material relating to §6, see

B. Schwartz, "An Analytic Method for the 'Difficult Crossing' Puzzles," *Mathematics, Magazine* **34** (1961) 187–193.

R. Bellman, "Dynamic Programming and Difficult Crossing Problems," *Mathematics Magazine* **35** (1962) 27–29.

S. Kumar, "The Crossing Puzzle and Dynamic Programming," *J. Indian Math. Soc.* **28** (1964) 169–174.

R. Fraley, K. L. Cooke, and P. Detrick, "Graphical Solution of Difficult Crossing Puzzles," *Mathematics Magazine* **39** (1966) 151–157.

For discussion of Wythoff's game see

H. S. M. Coxeter, "The Golden Section, Pyllotaxis and Wythoff's Game," *Scripta Math.* **19** (1953) 135–143.

J. C. Holladay, "Some Generalizations of Wythoff's Game and Other Related Games," *Mathematics Magazine* **41** (1968) 7–13.

Chapter   Eight

THE "TRAVELLING SALESMAN" AND OTHER SCHEDULING PROBLEMS

1. Introduction

So far we have considered certain classes of combinatorial problems which can be considered to belong to a new domain of mathematics called *scheduling theory*. The opening four Chapters were devoted to a routing problem, considered in various ways and with varying degrees of sophistication. The next three were devoted to two other types of scheduling processes appearing in the guise of familiar puzzles. The power of an abstract approach was illustrated by showing that these two latter problems, so different in form, were actually particular versions of the routing problem.

In this Chapter we wish to consider some additional scheduling questions. They belong to the same general class; yet, appear to be very much more difficult to resolve. As usual, there is little correlation between the simplicity of a statement of a problem and the complexity of its solution. In turn, we will examine the "travelling salesman" problem and the assignment problem. Both of these can be treated to some extent by the recurrence relation technique we have repeatedly employed. Following this, we pose a problem which appears completely to escape this general method. In the exercises, we will encounter the seven bridges of Konigsberg once again and the famous four color problem for maps, as well as a number of other questions.

As in the previous chapters, we will begin with word problems and convert them step-by-step into problems involving equations.

2.   A Mathematician's Holiday

It is Saturday afternoon, and we decide to go to the office and do some work, undisturbed by such distractions of the work week as telephones, visitors, and so on—provided, of course, that there is nothing much doing at the tennis court. All goes according to plan, until, racket in hand, we are at the door. Then:

"Dear," says Nina, "would you mind bringing some of Gloria's pots and pans back on your way? And would you stop at the store and pick up some Chinese food, and also some soda—and would you mail these letters at the Post Office, some need airmail stamps?"

"Dad," says Kirstie, "would you drive me over to Barbara's house— and pick up Linda on the way?"

"Hey, Dad," says Eric, "stop by the garage and see if my car is ready yet and pick up my laundry—and could you see if the book store has the book I ordered."

The task of performing these errands in an efficient fashion has a certain mathematical air about it. Could it be that we can use our previous techniques to tackle this question? Clearly, it is far more entertaining to think about how to solve the problem than to carry out the actual errands. We reluctantly put down our tennis racket and we spend the rest of the afternoon in the following fashion.

3.   A Map and a Matrix

To begin with, let us refer to a map of Santa Monica and environs, and locate the various destinations as indicated in Fig. 1. We then replace

TABLE 1

	1	2	3	4	5	6	7	8	9	10	11	12
1	—	$\frac{1}{4}$	2	$\frac{1}{4}$	$\frac{1}{4}$	1	$1\frac{1}{4}$	2	$1\frac{1}{2}$	$1\frac{1}{4}$	1	2
2	$\frac{1}{4}$	—	$1\frac{3}{4}$	$\frac{1}{2}$	$\frac{1}{4}$	$\frac{1}{2}$	$\frac{3}{4}$	$1\frac{1}{2}$	$\frac{1}{2}$	$\frac{1}{2}$	$\frac{3}{4}$	2
3	2	$1\frac{3}{4}$	—									
4	$\frac{1}{4}$	$\frac{1}{2}$		—								
5	$\frac{1}{4}$	$\frac{1}{4}$			—							
6	1	$\frac{1}{2}$				—						
7	$1\frac{1}{4}$	$\frac{3}{4}$					—					
8	2	$1\frac{1}{2}$						—				
9	$1\frac{1}{2}$	$\frac{1}{2}$							—			
10	$1\frac{1}{4}$	$\frac{1}{2}$								—		
11	1	$\frac{3}{4}$									—	
12	2	2										—

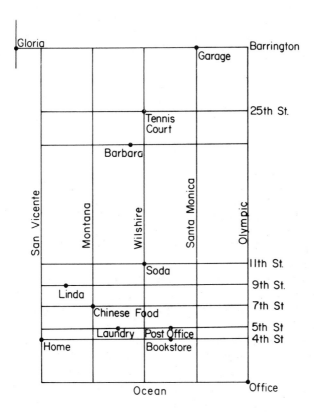

Fig. 1

Fig. 2

this figure by a set of points and a set of associated distances (see Fig. 2). Examining the map, we can write down the array of distances between these points (see Table 1). We suspect from our previous experience that this reduction to a set of points with an associated matrix will greatly facilitate our analytic formulation.

Exercise

1. Fill in the missing distances.

4. Enumeration

Let us now consider the problem of starting at 1 (home), visiting each of the numbered sites once and only once, and returning to 1, in such a way as to minimize the total distance travelled. As we have commented before, in many situations this not is at all the same as minimizing the total time required for the trip.

The first approach to be considered is, of course, that of enumeration of all possible paths. Some simple calculations which we carry out below show that this is not a feasible approach as soon as we have even a moderate number of errands to run. As in the routing problem, the totality of possibilities increases in an alarming fashion with the number of places to visit. We are, in effect, competing with $N!$, and this is a very uneven contest, as we already know. Let us go over some of the figures.

In the problem just described, there are 11 different places to visit, exclusive of home. Starting from home, we can go to any of 11 places first, any of 10 places next, any of 9 places after that, and so on. It follows that there are

$$11! = 11 \times 10 \times 9 \times \cdots \times 2 = 39,916,800 \qquad (1)$$

different paths. The proliferation of possible paths is quite remarkable.

Numbers such as 100!, 1000!, associated in this fashion with relatively moderately sized maps are unimaginably large. We can agree then that as far as we can see into the future, regardless of the kind of computer available, combinatorial problems of any significance cannot be resolved by brute force. Recall the discussion in § 15 of Chapter One. Let us turn then to the use of some mathematical techniques.

5. A Simplified Version

Suppose that we agree to keep Gloria's pots and pans for a while, let Barbara pick up Kirstie and Linda, ignore the Chinese food, the garage, the laundry, and the Post Office, and concentrate on essentials:

(a)   Tennis court (7)
(b)   Soda (9)
(c)   Bookstore (11)
(d)   Office (12)

What is the optimal path now? Let us draw some trees (see Fig. 3).

We observe that the problem is similar in many ways to that of routing considered in the first Chapter. There is, however, one most significant difference. Since we now insist upon visiting all of the vertices, in order to determine how to continue optimally at any stage, we must know either where we have already been, or where we have not yet been. As we shall see, this seriously complicates the numerical solution. The "state" at any stage is now a much more complex concept.

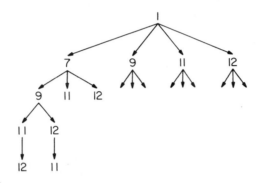

Fig. 3

6. Recurrence Relations

As before, let us introduce some quantities whose determination will let us obtain the minimal path. Let

$$f_7 = \text{the minimum distance to travel, starting at 7 and}$$
$$\text{going to the remaining points 9, 11, 12, in some}$$
$$\text{order, and then 1} \tag{1}$$

and let quantities f_9, f_{11}, f_{12} be defined similarly.

Since we must go somewhere from 1, we see that the overall minimum distance is given by the quantity

$$\min [d_{1,7} + f_7, d_{1,9} + f_9, d_{1,11} + f_{11}, d_{1,12} + f_{12}] \tag{2}$$

Similarly, let

$f_{7,9}$ = the mininum distance to travel, having been at 7,
continuing to 9, and ending at 1   (3)

and let the quantities $f_{7,11}$, $f_{7,12}$, $f_{9,7}$, $f_{9,11}$, $f_{9,12}$, etc., be defined in analogous fashion. Then, reasoning as before, we see that

$$f_7 = \min [d_{7,9} + f_{7,9}, d_{7,11} + f_{7,11}, d_{7,12} + f_{7,12}]   (4)$$

where the distances $d_{7,9}$, $d_{7,11}$, $d_{7,12}$ and so on are readily obtained from our master array (Table 1).

To determine quantities such as $f_{7,9}$ we introduce the quantities $f_{7,9,11}$, $f_{7,9,12}$, and so on. These quantities are immediately determined, since

$$f_{7,9,11} = d_{11,12} + d_{12,1}$$
$$f_{7,9,12} = d_{12,11} + d_{11,1}   (5)$$

In other words, once at 11 or 12, and having been at 7 and 9, 12 or 11 respectively, we go directly home.

Exercise

1. Determine the optimal path or paths, and the quantities f_7, f_9, f_{11}, f_{12}, $f_{7,11}$, etc.

7. Do All the Errands!

Suppose that the family presents a united front and indignantly insists that the price of peaceful tennis, or, more precisely, peaceful after-tennis, is the carrying out of *all* of the errands. Let us see what this entails.

Following the path laid out in the preceding section, we can transform the problem into a routing problem of considerably larger dimension by observing that the state at any stage can be defined as the set of points already visited, or, equivalently, the set of states yet to visit. Thus, if we are at i_k having been at i_1, i_2, $\cdots$, i_{k-1}, we introduce the generalized state $(i_1, i_2, \cdots, i_k)$. The effect of going to the next point i_{k+1} is to convert the previous state into the new state $(i_1, i_2, \cdots i_{k+1})$. An interesting point to note is that once at i_{k+1}, and concerned only with an optimal continuation, it makes no difference in what order we visited the previous points.

Let us then introduce the functions

$f(i_1, i_2, \cdots, i_k)$ = the minimum remaining distance to travel,
starting at i_k with the knowledge that we
have already visited i_1, i_2, $\cdots$, i_{k-1} in some
order, where $2 \leq i_1, i_2, \cdots, i_k \leq 12$ (1)

We have changed over from the use of subscripts to the use of function arguments for the sake of the typographer.

Then we readily obtain the recurrence relation

$$f(i_1, i_2, \cdots, i_k) = \min_j [d(i_k, j) + f(i_1, i_2, \cdots, i_k, j)] \quad (2)$$

Here $d(i_k, j)$ is the distance from i_k to j. As above, the quantities $f(i_1, i_2, \cdots, i_{10})$ are immediately determined by the relation

$$f(i_1, i_2, \cdots, i_{10}) = d(i_{10}, i_{11}) + d(i_{11}, 1) \quad (3)$$

In other words, we must go to the remaining place and then home.

8. Feasibility

Is this really a reduction of the original enumeration process, or have we merely systematized it? It is indeed a decrease in the inspection of possibilities. The reason is first that when we are at i_k, we need to know only where we have been, but not the order in which we carried out the preceding errands, and secondly that only optimal continuations are examined.

Let us determine the number of possible sets of $k - 1$ different numbers $(i_1, i_2, \cdots, i_{k-1})$ chosen from the ten integers that remain when we delete i_k from the numbers 2 through 12. It is the binomial coefficient

$$\frac{10!}{(k - 1)! \ (11 - k)!} \quad (1)$$

which assumes the following values as k goes from 1 to 11,

$$1, 10, 45, 120, 210, 252, 210, 120, 45, 10, 1 \quad (2)$$

Since i_k can have any of 11 different values, we see that we have at most at any stage

$$11 \times 252 = 2772 \quad (3)$$

sets of vales $\{(i_1, i_2, \cdots, i_k)\}$ with which to contend. This hardly strains our rapid-access storage capability.

When there are 5 points remaining we can switch over to keeping track of where we have yet to go, rather than where we have been.

9. How Many Errands Could We Do?

How many places could we visit—in an efficient fashion, that is? In other words, how large a problem of this nature can we solve? We see that it depends on the quantity

$$\frac{N(N-1)!}{k!(N-1-k)!} \tag{1}$$

where $1 \le k \le N-1$. It is easy to see that if N is even, the middle binomial coefficient is the largest. If N is odd, the middle two binomial coefficients are largest. Take N odd, say $N = 2M + 1$, so that the binomial coefficient is largest when $k = M$. Then the quantity of interest is

$$(2M + 1)\left(\frac{(2M)!}{(M!)^2}\right) \tag{2}$$

With present day computers we begin to encounter basic capacity constraints at about $M = 12$. If, however, we wished to use disc storage, we could increase the size of the problem.

10. Trading Time for Storage

The problem we have just considered is often called the "travelling salesman" problem. A familiar version is the following: "What path does a salesman pursue in order to visit each of the state capitals once and only once in the shortest time?" This type of problem of determining minimal closed circuits had earlier been considered by W. R. Hamilton in connection with certain three dimensional geometric figures.

It is easy to see that problems of this general nature enter into many economic and industrial activities. Consequently, a great deal of effort has been expended trying to obtain effective solutions. Sometimes, the limitation in computer capability can be circumvented by trading time for storage. If the problem is important enough, we may be willing to expend great amounts of time in its solution. This would be the case, for example, if we were attempting to determine a route which would be followed day in and day out, year after year.

One way to effect this trade is to combine enumeration with the analytic solution described above. Suppose, for example, that there are 31 points. Starting at 1 we can go to four other points in $30 \times 29 \times 28 \times 27$ different ways. Let us say the path is $[1, i_1, i_2, i_3, i_4]$. Starting at i_4 we must traverse the remaining points in minimum time, or minimum distance, and return to 1. Each of these problems can be resolved. Thus, we can solve the original problem involving 31 points at the expense of determining 657,720 numbers obtained by combining the answers to the 27 point problem and the distance already traversed.

This number, 657,720, can be considerably reduced by observing that having arrived at i_4 it makes no difference as far as the remaining 27 point problem is concerned as to how we traversed the points $[1, i_1, i_2, i_3]$. In particular, it is seen that of all ways we could start at 1 and end at i_4, passing through i_1, i_2, i_3, we want to keep only the path of minimum distance. Thus, since i_4 can be chosen in 30 ways, we see that there are $30 \times 29 \times 28 \times 27/6$ possibilities. This reduces the previous number to 109,620. This is still considerable. However, it is interesting to note that the 27 point problems could be done simultaneously on different computers, thus vastly reducing the time required to obtain an answer, provided that one was willing to cover the cost of calculation.*

In general, the problem is vastly simpler due to the fact that it is only possible or sensible to go to one of a few nearby points. Thus, if we allow only four "nearest neighbors," the number of 27 point problems is reduced to 4^5 which is $2^{10} = 1024$, a much more reasonable quantity.

The point of all of this is that in any particular scheduling process we can always combine a general approach with intrinsic structural properties, common sense, experience and ingenuity, to obtain feasible approaches.

11. How to Build a Tennis Court

Let us now consider a rising young executive, facing the constant task of assigning groups of men to perform certain jobs. In particular, suppose that his position is with a construction company and the job is that of building a tennis court. He has a crew of seven, Bill, Carl, Harry, Joe, Mac, Ray, and Walt, each of whom, with proper patience and care, can be used in different ways. There are actually seven categories of labor: Foreman, Blueprint Reader, Bulldozer, Line-Painter, Net-man, Book-keeper, and Leveller. How does the executive determine who is to do what?

As is to be expected, some men are better at some jobs than others. On the basis of long experience with the construction of tennis courts, the executive has learned the cash value of assigning a particular man to a

TABLE 2

	Foreman	Blueprint	Bulldozer	Lines	Net Man	Bookkeeper	Leveller
Bill		10		1	5		5
Carl	10		10	1		10	1
Harry		10		1	10	5	5
Joe	10	1	10	1	10	1	5
Mac		10	1	1	5	10	5
Ray		10	10	1	5	1	10
Walt	10	1	1	1	1	10	5

* This is an important point in connection with the contemporary development of parallelization in computers.

particular job. To guide his decision-making, he has constructed the array in Table 2.

The empty spaces are to be filled in with zeroes.

The executive realizes that none of his workers are particularly expert at lines. However, their other abilities more than compensate for this deficiency. How does he assign the men to the jobs so as to maximize his total profit?

12. Sequential Selection

In place of using names for both the men and the tasks, let us use numbers. Let $i = 1, 2, \cdots, 7$ denote the men and $j = 1, 2, \cdots, 7$ denote the tasks. We then obtain the matrix in Table 3.

TABLE 3

$$\begin{bmatrix} 0 & 10 & 0 & 1 & 5 & 0 & 5 \\ 10 & 0 & 10 & 1 & 0 & 0 & 1 \\ 0 & 10 & 0 & 1 & 10 & 5 & 5 \\ 10 & 1 & 10 & 1 & 10 & 1 & 5 \\ 0 & 10 & 1 & 1 & 5 & 10 & 5 \\ 0 & 10 & 10 & 1 & 5 & 1 & 10 \\ 10 & 1 & 1 & 1 & 1 & 10 & 5 \end{bmatrix}$$

The ijth element in this matrix is the value of assigning the ith man to the jth job.

Let us think of the process of assignment as a sequential process in which we first assign one of the seven men to the first job, then one of the six remaining to the second job and so on. We see that once again we are tracing a path through a network with the constraint that we must keep in mind where we have been. Let us enumerate the jobs in some fixed order. Then, accordingly, one of the seven men must be assigned to the first job. Let

$f(k) =$ the maximum profit obtained assigning the remaining six men to the remaining six jobs, having assigned the kth man to the first job, $k = 1, 2, \cdots, 7$ (1)

The required maximum profit is

$$\max_k [f(k)] = \max [f(1), f(2), \cdots, f(7)] \qquad (2)$$

Similarly, let us introduce the quantities

$f(k_1, k_2) = $ the maximum profit obtained assigning the k_1st man to the first job, the k_2nd man to the second job and assigning the remaining five men to the remaining five jobs in an optimal fashion   (3)

Then our familiar line of reasoning yields the relations

$$f(k) = \max_{l \neq k} [a(k, l) + f(k, l)], \quad k = 1, 2, \cdots, 7 \quad (4)$$

The numbers $a(k, l)$ are obtained from the table given above. Similarly, we obtain a relation connecting $f(k, l)$ with $f(k, l, m)$,

$$f(k, l) = \max_{r \neq k, l} [a(l, 2) + f(k, l, r)] \quad (5)$$

and so on.

13. Feasibility

How many different values must we store at each stage? There are seven values of $f(k)$; $7 \times 6 = 42$ different values of $f(k_1, k_2)$; $7 \times 6 \times 5 = 210$ different values of $f(k_1, k_2, k_3)$; $7 \times 6 \times 5 \times 4 = 840$ different values of $f(k_1, k_2, k_3, k_4)$; $7 \times 6 \times 5 \times 4 \times 3 = 2520$ different values of $f(k_1, k_2, k_3, k_4, k_5)$; and, finally, $7! = 5040$ different values of $f(k_1, k_2, k_3, k_4, k_5, k_6)$.

Now, of course, we don't have to use recurrence relations to determine $f(k_1, k_2, k_3, k_4)$. Since there are only three men to assign to the remaining three jobs, a total of six possibilities, it is quite easy to obtain the values of this function for any given set of $k_1, k_2, k_3,$ and k_4. Having determined these 840 values, the set of numbers $f(k_1, k_2, k_3, k_4)$, we then use recurrence relations to obtain the 210 values of $f(k_1, k_2, k_3)$, then the 42 values of $f(k_1, k_2)$, the seven values of $f(k_1)$, and, finally, the desired maximum value.

It is clear that if the number of different jobs is large, we are going to run into difficulties with the foregoing approach, namely the usual difficulties associated with N! There are several factors which, in this case, provide some relief from the overwhelming magnitude of N! In the first place, there are other approaches to this particular problem which are very much more efficient. We will discuss this briefly below in § 16. Secondly, in practice, simpler and more manageable versions of the assignment problem occur. We discuss one aspect of this in the following sections.

Exercises

1.   Joe decides to be foreman, and Carl insists upon doing the bull-dozing. Determine, using recurrence relations of the type given above, how

the other men should be assigned in order to maximize the profit.

2. Determine the solution to the original problem.

14. Simplified and More Realistic Assignment Problem

In actuality, it seldom happens that every man can carry out every job. What simplifies the assignment task in most cases is that each man can only be used in a small number of positions.

Suppose in the present case that each man can be used in three positions as indicated in the following table.

TABLE 4

Foreman	Blueprint	Bulldozer	Lines	Net Man	Bookkeeper	Leveller
Bill		10		5		5
Carl	10		10		10	
Harry		10	1	10		
Joe	10		10	10		
Mac		10		5		5
Ray			10		1	10
Walt	10			1		5

We can, if we wish, fill in the missing boxes with zeroes, and so reduce the problem to the more general problem treated first. We would expect, however, that exploitation of this special structure will considerably reduce the rapid-access storage problem.

Rather than $7! = 5040$ possibilities, we now have at most $3^7 = 2187$. This is not a particularly impressive reduction; but for ten jobs, we would compare $10! = 3,628,800$ with at most $3^{10} = 59,049$. Furthermore, we can partition the problem in the manner discussed in § 10 and thus treat problems of some magnitude in this fashion. Observe how the set of possible assignments simplifies.

If Bill is assigned as foreman, then Carl, Harry, or Joe can fill the second job of blueprint reader. But if Carl is made foreman, then either Harry or Joe *must* be blueprint reader.

Thus, at the first stage, we have only the functions $f(1)$, $f(2)$, $f(3)$ to consider and the desired maximum profit is

$$\max [f(1), f(2), f(3)] \qquad (1)$$

At the second stage, we need consider

$$[f(1, 2), f(1, 3), f(1, 4), f(2, 3), f(2, 4), f(3, 2), f(3, 4)] \qquad (2)$$

We see that the number of possible assignments is far less than the $3^7 = 2187$ upper bound mentioned above.

15. Constraints

Another realistic feature that considerably simplifies the assigning of men to jobs is that certain groups of men either must work together, or cannot work together. This means that once a few key men have been assigned, the other assignments are completely determined. Let us consider an example of this, in terms of the original tennis-court problem.

Suppose that only Joe or Carl can be foreman, and that Joe insists upon having Bill as blueprint reader while Carl demands Harry if he is chosen. Furthermore, Harry will agree to be blueprint reader only if Mac does the bookkeeping and Ray does the levelling. The choice of blueprint reading is between Harry and Joe.

Assignments are quite limited by these conditions. For the first job, we have only two candidates, No. 2 and No. 4. If 2 is put in the first job, then 3 must go into the second job; if 4 goes into the first job, then 1 must go into the second job.

Continuing in this way, we see that the number of possible work crews is quite limited. In larger problems, structural peculiarities of this nature are usually present. If exploited, the magnitude of the problems can be greatly reduced. How to spot these features is at the present time an art which can be improved with experience. A systematic exploration of problems of this nature seems to present a serious mathematical challenge. It is to be expected that many significant scheduling problems can be profitably attacked using both logical and topological techniques.

16. Discussion

We considered in the foregoing sections two scheduling problems. In both cases we rather quickly stubbed our toes against the barrier of $N!$. Now the interesting fact is that in the first problem, the travelling salesman problem, no better technique exists today for producing a guaranteed solution. There are, however, available methods which have a high probability of solving the problem.* In the second case, the assignment problem, there exist elegant and powerful mathematical methods which can solve problems of this nature involving hundreds of men and jobs in a matter of minutes.

This is part of the fascination of this field of mathematics. The most

---

* We are thinking of "branch and bound" techniques.

innocent appearing problem can lead to incredible difficulties; a seemingly involved problem can yield to a simple method.

Let us note in passing that the assignment problem turns out to be equivalent to the following problem: Given a set of numbers a_{ij}, $i, j = 1$, $2, \cdots, N$, determine the quantities x_i, $i = 1, 2, \cdots, N$ such that we have

$$a_{ij} \leq x_i + x_j \qquad i, j = 1, 2, \cdots, N \qquad (1)$$

with $\sum_{i=1}^{N} x_i$ a minimum. In this formulation, it appears easier to obtain approximate solutions.

17. 2^N

To illustrate the statement made above that innocuous problems can lead to amazing difficulties, consider the question of evaluating the power 2^N in an efficient fashion. At first glance there seems to be little of interest here. We compute

$$2^2, 2^3, \cdots, 2^{N-1}, 2^N \qquad (1)$$

multiplying by 2 at each step. Recall, however, one of the perils of multiplication: round-off error. Second, recall that multiplications require a considerable quantity of time, compared to additions, for example. From both points of view, there is considerable merit in planning the calculation of 2^N in such a way as to reduce the total number of multiplications.

In some cases, it is clear that we can greatly improve on the foregoing procedure which requires $(N - 1)$ multiplications. Consider, for example, the calculation of 2^{64}. We can proceed as follows

$$2 \times 2 = 2^2, \; 2^2 \times 2^2 = 2^4, \; 2^4 \times 2^4 = 2^8, \; 2^8 \times 2^8 = 2^{16},$$
$$2^{16} \times 2^{16} = 2^{32}, \; 2^{32} \times 2^{32} = 2^{64} \qquad (2)$$

a total of 6 multiplications. This is a considerable improvement over 63.

Consider the computation of 2^{70}. We could proceed as before to calculate 2^{64}, then calculate $2^{68} = 2^{64} \times 2^4$, then $2^{70} = 2^{68} \times 2^2$. This requires a total of 8 multiplications. Or we could calculate 2^{32} as above, then $2^{35} = 2^{32} \times 2^3$ then $2^{70} = 2^{35} \times 2^{35}$. This requires 7 multiplications. In general, how does one calculate 2^N using the least number of multiplications? Surprisingly, nobody seems to know the answer for general values of N.

18. A Branching Process

Let us see what happens when we apply the foregoing techniques. Since the base remains the same, we can concentrate upon the exponents and speak purely in terms of addition. At the first stage, the only operation we can perform is that of adding one to itself, producing the pair (1, 2). At the second stage, we can add 1 to 2 or 2 to itself. At the next stage the possibilities are shown by the following figure.

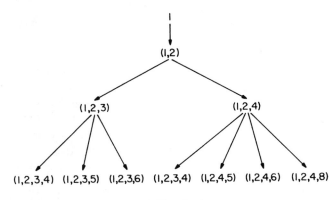

Fig. 4

Thus, at each stage we have a set of exponents corresponding to powers of 2 already computed. A choice of a multiplication to perform is equivalent to a choice of an addition of two of the numbers in the exponent set.

19. Recurrence Relations

Let us attempt to apply the method of recurrence relations to this problem. To this end, we can introduce the function

$$f(k_1, k_2, \cdots, k_r) = \text{the minimum number of remaining}$$
additions to obtain N, starting with
the exponent set $(k_1, k_2, \cdots, k_r)$ where
$k_1 = 1, k_2 = 2$ and $3 \leq k_3 < k_4 < \cdots$
$< k_r < N$ (1)

We can readily write an equation connecting $f(k_1, k_2, \cdots, k_r)$ with $f(k_1, k_2, \cdots, k_r, k_i + k_j)$, namely

$$f(k_1, k_2, \cdots, k_r) = 1 + \min_{\{k_i, k_j\}} f(k_1, k_2, \cdots, k_r, k_i + k_j) \qquad (2)$$

The difficulty is that this ceases to be a feasible algorithm even for relatively small values of N.

Exercises

1. Calculate $f(1, 2)$ for the cases where $N = 10, 15, 20$.

2. Assuming a rapid access storage of 64,000, how large a value of N can be taken?

3. Assuming the use of a slow access storage of 8×10^6, how large a value of N is permissible?

20. Discussion

The motivation behind the choice of the foregoing was a desire to show how easily difficult mathematical problems arise from simple questions that begin, "How do we $\cdots$". Some of these problems, as we saw, can be formulated as abstract routing problems and approached by means of recurrence relations. In most cases, this is neither a computationally feasible approach nor particulary elegant analytically. In some cases, as we noted in the 2^N-problem, we know of no algorithm for obtaining the solution which is significantly better than enumeration of possibilities.

A great deal of effort has been expended in this area using "heuristic programming." By this we mean a use of approximate policies combined with computer experimentation and on-line modification. All of this overlaps into the new and vital area of artificial intelligence, one of the most exciting and challenging of contemporary intellectual activities.

A number of references for further reading will be found at the end of the chapter for those who want to hunt really big game. Some may be attracted by the intellectual challenge, some by the importance of the applications. Some may wish to explore the possibilities of analytic and topological approaches, some may want to test the powers of the much-vaunted computer. Whatever the motivation, this new field of combinatorics and scheduling represents Opportunities Unlimited. Fortunate indeed is the young mathematician of today!

Miscellaneous Exercises

1. Suppose that we are given a particular map, such as that appearing in Fig. 5 below, and are asked to determine whether or not we can color the map with four distinct colors in such a way that no two contiguous states are of the same color. The heavy lines, five of them in this case, are called "separating boundaries." They have the property that the subregion contained between the ith and $(i + 1)$st separating boundaries separate the entire map into two parts which have no common boundary.

   In place of the original problem, consider the problem of minimizing the total number of "overlaps." An overlap occurs when two contiguous states have the same color. Let

$f_i(c) =$ the number of overlaps incurred coloring the region to the left of boundary (i) using an optimal coloring scheme, given that the strip between $(i-1)$ and i is colored with a color scheme C

Let

$F_i(C, C') =$ number of overlaps resulting from a coloring scheme C' in the region between (i) and $(i+1)$ when the region between (i) and $(i-1)$ has the scheme C

Show that

$$f_i(C) = \min_{C'} [F_i(C, C') + f_{i+1}(C')] \qquad i = 1, 2, 3, 4$$

with $f_5(C) = \min_{C'} F_5(C, C')$

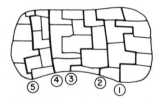

Fig. 5

2. The best color scheme for the original map is determined by minimizing $f_1(C)$ over C. This is carried out by straightforward enumeration. Show that a subregion with s states can be colored in k^s different ways, allowing k different colors. Consider a map which possesses separating boundaries with the property that no subregion between (i) and $(i+1)$ possesses more than s states. Show that the four-color conjecture is true for $s = 1, 2$.

3. Using a digital computer, show that the four-color conjecture is true for $s = 3, 4, 5$, for any finite number of separating boundaries.

4. Calculate the rapid access storage requirements required to go up to $s = 10$. What can be done with current disk storage?

5. Is the four-color conjecture valid for a map of the United States? (The answer is obviously "yes" as a glance at any atlas will show. But assume that you are starting with an uncolored map.)

The foregoing appeared in

R. Bellman, "An Application of Dynamic Programming to the Coloring of Maps," *I.C.C. Bulletin* **4** (1965) 3–6.

6. Let us consider the Konigsberg Bridges once again. This time we ask for a path which traverses every edge at least once, and for which the number of repetitions of edges is a minimum. Show that the tracing of a path can be considered to be a process of moving through a succession of states where a state is a list $[Q, e_1, e_2, \cdots, e_k]$, with Q the node at which the tracing point lies at the moment and $[e_1, e_2 \cdots, e_k]$ the set of edges still to be traversed. Here $1 \leq k \leq M$ where M is the total number of edges. Why in this case don't we wish to list the edges already traversed?

7. Assign a length of one to each edge and introduce the function $f(Q;$ $e_1, e_2, \cdots, e_k)$ = length of shortest path starting at the point Q and including all edges $e_1, e_2, \cdots, e_k$ at least once. Show that

$f(Q; e_1, e_2, \cdots, e_M) = M +$ the minimal number of edge repetitions for a path starting at Q and traversing every edge at least once

and that

$C = \min\limits_{Q} f(Q; e_1, e_2, \cdots, e_M) - M =$ the minimal number of edge repetitions for a path in the graph which traverses every edge at least once.

8. What modifications are required if we wish to determine a path that returns to its starting point?

9. Let us define

$S(Q)$ = set of nodes Q' such that QQ' is an edge; i.e., $S(Q)$ is the set of nodes accessible from Q

Let E denote any set of edges and write $f(Q; E)$ in place of $f(Q; e_1,$ $e_2, \cdots, e_k)$. Then

$$f(Q; E) = \min\limits_{[Q_1Q_2]} [1 + f(Q_1; E), 1 + f(Q_2; E_2]$$

where the minimization is over Q_1 and Q_2 in $S(Q)$ such that QQ_1 is not a member of E, QQ_2 is a member of E, $E_2 = E - \{QQ_2\}$.

10. Show that the minimal number of edge repetitions can be computed in the following sequential procedure: Determine the pairs (Q, e_i) for which $f(Q; e_i) = 1$; determine the pairs (Q, e_i) for which $f(Q, e_i) = 2$ and the (Q, e_i, e_j) for which $f(Q, e_i, e_j) = 2$, using the foregoing equation, and so on.

For the foregoing, see

K. L. Cooke and R. Bellman, "The Konigsberg Bridge Problem Generalized," *Jour. Math. Anal. Appl.* (forthcoming).

11. Ten coins are placed in a row. A coin may be moved over two of those adjacent to it onto the coin next beyond them. It is required to move the counters according to these rules so as to arrange them in five equidistant couples. How do we proceed?

12. If two superimposed counters are considered as only one, what are the solutions?

W. W. Rouse Ball, *Mathematical Recreations and Essays*, Macmillan Company, New York, 1947, 119–120.

13. Place four white checkers and four black checkers alternately in a line in contact with one another. Make four moves, each of a pair of two contiguous pieces, without altering the relative position of the pair, to form a continuous line of four black checkers followed by four white checkers.

Ibid., 121.

14. A heavy motor vehicle reaches the edge of a desert 400 miles wide. The vehicle averages only one mile to the gallon of gas, and the total available gasoline tank capacity, including extra cans, is 180 gallons, so it is apparent that gasoline dumps will have to be established in the desert. There is ample gas to be had at the desert edge. With wise planning of the operation, what is the least gas consumption necessary to get the vehicle across the desert?

L. A. Graham, *Ingenious Mathematical Problems and Methods*, Dover Publications, Inc., New York, 1959, 10.

15. A rookie electrician was ordered by his foreman to row across a stream with a bundle of 21 wires and connect them to a control board on the other side. The poor novice got back to his boss and admitted shamefacedly that he hadn't labelled the wires. "Get going," shouted the foreman, "and label those wires by yourself with the least rowing and use no unnecessary instruments, or you are fired." How did he do it?

Ibid., 11.

16. A tourist wants to visit *N* different sites of historical and artistic interest during his stay in Rome, subject to the constraint that he visit no more than *M* a day. How does he plan his sightseeing?
    See

R. Bellman, "An Algorithm for the Tourist Problem," (to appear).

17. *The Underground Maze* "The only way out of the yard that I now was in was to descend a few stairs that led up into the centre (A) of an underground maze, through the winding of which I must pass before I could take leave by the door (B). But I knew full well that

in the great darkness of this dreadful place I might well wander for hours and yet return to the place from which I set. How was I then

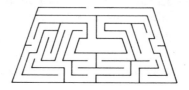

to reach the door with certainty? With a plan of the maze it is but a simple matter to trace out the route, but how was the way to be found in the place itself in utter darkness?"

H. E. Dudeney, *The Canterbury Puzzles*, Dover Publications, Inc., New York, 1958, 79–80.

18. Write 31 using the digit 3 five times and using only the ordinary arithmetic operations plus juxtaposition and exponents.

19. *The Battle of Numbers* This is a game for two players A and B. A starts and selects an integer greater than zero and less than or equal to 10. B adds to this an integer not greater than 10. The first to reach 100 wins. How do they play?

M. Kraitchik, *Mathematical Recreations*, Dover Publications, Inc., New York, 1942, 83.

20. A smuggler wishes to cross just once every frontier of each of the various countries of Europe. What is his route, if he agrees to take his chances and travel in such a way as to minimize the total number of frontiers crossed more than once?

21. How shall we place in line 3 hunters, 3 wolves, 3 goats and 3 cabbages without disturbing the peace by having a hunter next to a wolf, a wolf next to a goat, or a goat next to a cabbage, and without creating rivalry by having 2 hunters, wolves, goats, or cabbages side by side? See

M. Kraitchik, *Mathematical Recreations*, Dover Publications, Inc., New York, 1942, Ch. 9.

22. How may k rooks be placed on a $k \times k$ chessboard so that no one can capture any of the others?

Ibid., Ch. 10.

23. How may k queens be placed on a $k \times k$ chessboard so that no one can capture any of the others? Show that there is no solution for $k \leq 4$, and determine the solutions for $5 \leq k \leq 8$.

Ibid.

24. Consider an $M \times N$ chessboard and the possibility of a knight being able to tour the board, going once and just once to each square. Show that
    a. If one dimension is less than 3, no tour is possible.
    b. If one dimension is 3, the other is at least 7.

25. N slips are placed in a box, each slip with a number written on it, and all the numbers different. A player does not know the numbers in advance, and draws one slip after another. He wins if he breaks off just after drawing the highest of the N numbers. How should he proceed? See

D. V. Lindley, "Dynamic Programming and Decision Theory," *Appl. Stat.* **10** (1961), 39–51.

S. Vajda, "The Marriage Problem, An Application of Dynamic Programming," *J.R.N.S.S.* **19** (1964).

26. A young lady decides that she wants to get married. The question is to whom. She expects to receive in the near future N proposals of marriage and she must make her decisions in such a way that she accepts the best available partner, who is not necessarily the first who proposes to her. How should she proceed?

27. Polyominoes are simple geometric figures of the following type:

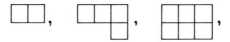

Fig. 7

and so on. How do we solve the problem of fitting a finite collection of such figures together to form a rectangular region such as

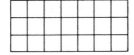

Fig. 8

For interesting accounts of some of the intriguing problems that arise from the consideration of these figures, see

S. W. Golomb, *Tiling with Polyominoes*, USCEE Report 155, University of Southern California, 1965.

S. W. Golomb, *Polyominoes*, Scribner's, New York, 1965.

S. W. Golomb, "Replicating Figures in the Plane," *Mathematical Gazette* **48** (1964) 403–412.

H. Wang, "Games, Logic, and Computers," *Scientific American*, **213** (1965) 98–106.

28. The polynominal $a_0 x^n + a_1 x^{n-1} + \cdots + a_{n-1} x + a_n$ can clearly be computed for any value of x and assigned values for the coefficients with the aid of $2n - 1$ multiplications. It can also be computed with the aid of only n multiplications by calculating $a_0 x$, adding a_1, calculating $(a_0 x + a_1)x$ etc. Can one do better? The problem has recently been solved and the answer is "no." See

V. Ya Pan, "Methods of Computing Values of Polynomials," *Russian Math. Surveys* **21** (1966) 105–136.

S. Winograd, "On the Number of Multiplications Required to Compute Certain Functions," *Proc. Nat. Acad. Sci.* **58** (1967) 1840–1842.

29. EXTENDAPAWN is played on a rectangular board consisting of 3 rows and n columns together.

    The game is begun with pawns of one kind, say X, filling the base row and pawns called 0 filling the top row, the middle row being empty.

    A move by a player consists of
    1. moving a pawn straight forward one square into an empty square, or
    2. "capturing" one of the opponent's pawns diagonally. The object of the game of n-columns is to win by:

    Type 1. moving into the opponent's row with one pawn, or
    Type 2. depriving the opponent of any further move (i.e., by moving last).

    How does one win?

    See

J. R. Brown, "EXTENDAPAWN—An Inductive Analysis," *Mathematics Magazine* **38** (1965) 286–299.

30. Examine the possibility of using recurrence relations to determine optimal play in checkers and king-pawn endings in chess.

    See

R. Bellman, "Dynamic Programming and Markovian Decision Processes with Particular Application to Baseball and Chess," *Applied Combinatorial Mathematics*, Wiley, New York, 1964, 221–236.

R. Bellman, "On the Application of Dynamic Programming to the Determination of Optimal Play in Chess and Checkers," *Proc. Nat. Acad. Sci.* **53** (1965) 244–247.

R. Bellman, "Stratification and Control of Large Systems with Applications to Chess and Checkers," *Information Sciences* **1** (1968) 7–21.

31. Let us consider the following problem: A group of cement factories produces several types of cement, but each factory produces only one type. There is also a group of purchasers and each purchaser may

need several types of cement. The amounts supplied and the demands are assumed to be known for each cement factory and each purchaser. Each cement factory has several trucks, but there is only one dispatcher for all of the trucks from all of the factories. It is assumed that the entire load of each truck is for one purchaser only. A truck begins its workday by leaving its base depot loaded and ends its day by returning to it empty. During the day it may be required to transport cement from any of the cement factories. The distances from various factories to individual purchasers are known. The problem to be solved is that of finding a truck schedule such that cement in the needed quantities is delivered daily to the individual purchasers and in such a manner that the total truck-kilometers travelled will be as small as possible.

W. Szwarc, "The Truck Assignment Problem," *Naval Res. Log.* Q. **14** (1967) 529–557.

32. What is the minimum number of weighings which suffice to determine the defective coin in a set of N coins of the same appearance, given an equal arm balance and the information that there is exactly one defective coin present which is lighter?

33. Suppose that we only know that it is of different weight. Do the same number of weighings suffice? See

J. M. Hammersley, "A Geometrical Illustration of a Principle of Experimental Directives," *Phil. Mag.* **39** (1948) 460–466.

34. Suppose that we know that there are two defective coins present. See

R. Bellman and B. Gluss, "On Various Versions of the Defective Coin Problem," *Information and Control*, **4** (1961), 118–131.

35. Determine the sequence of n items that are processed on 2 machines which minimizes the total time, where the processing requires that the machines be used in the same numerical order for any item and that the items be sequenced identically on each machine.

S. M. Johnson, "Optimal Two and Three Stage Production Schedules with Set-up Times Included," *Naval Res. Log. Quart.* **1** (1954), 61–68.

I. Nabeshima, "Computational Solution to the M-Machine Scheduling Problem," *J. Oper. Res. Soc. Japan* **7** (1965) 93–103.

I. Nabeshima, "Branch and Bound Algorithms for Optimal Sequencing of Jobs Through Machines where Passing Is Allowed," Reports of U. of Electro-Communications, No. 21 (1966) 63–73.

I. Nabeshima, "Some Extensions of the M-machine Scheduling Problem," *J. Oper. Res. Soc. Japan* **10** (1967), 1–17.

36. Consider the six permutations

$$0011, \quad 0101, \quad 0110, \quad 1001, \quad 1010, \quad 1100$$

Is it possible to arrange these in a sequence such that each is derived from its predecessor by the interchange of a pair of adjacent digits? What if a permutation can be repeated once? See

D. H. Lehmer, "Permutation by Adjacent Interchanges," *Amer. Math. Monthly* **72** (1965), no. 2 part II, 36–46.

Bibliography and Comment

For an elementary introduction to scheduling problems, see

A. Kaufmann and R. Faure, *Introduction to Operations Research*, Academic Press, New York, 1968.

A. Kaufman, *Graphs, Dynamic Programming and Finite Games*, Academic Press, New York, 1967.

For an advanced discussion, see

L. R. Ford and D. R. Fulkerson, *Flows In Networks*, Princeton University Press, 1962.

Treatments of the assignment and travelling salesman problems will be found in these volumes. See also

R. Bellman, "Dynamic Programming Treatment of the Travelling Salesman Problem," *J. Assoc. Comp. Mach.* **16** (1961) 61–63.

D. R. Fulkerson, "Flow Networks and Combinatorial Operations Research," *Amer. Math. Monthly* **73** (1966) 115–137.

R. E. Gomory, "The Travelling Salesman Problem," *Proc. IBM Sci. Computing Symp. Combinatorial Problems*, IBM (1966).

A powerful method for treating scheduling problems is the "branch and bound" method. See

E. L. Lawler and D. E. Wood, "Branch-and-Bound Methods: A Survey," *Oper. Research* **14** (1966) 699–719.

M. Bellmore and G. L. Nemhauser, "The Travelling Salesman Problem: A Survey," *Oper. Research* **16** (1968) 538–558.

J. M. Dobbie, "A Survey of Search Theory," *Oper. Research* **16** (1968) 525–537.

C. E. Nugent, T. E. Vollmann and J. Ruml, "An Experimental Comparison of Techniques for the Assignment of Facilities to Locations," *JORSA* **16** (1968) 150–173.

S. Gaunt, "A Non-Computer Method Using Search for Resolving the Travelling Salesman Problem," *CORS J.* **6** (1968) 44–54.

J. M. Hammersley, "Monte Carlo Methods for Solving Multivariable Problems," *Annals N. Y. Acad. Sci.* **86** (1960) 844–874.

An interesting paper is

A. K. Obruča, "Spanning Tree Manipulation and the Travelling Salesman Problems," *Comuter J.* **10** (1968) 374–377.

In this paper the author starts with a spanning subtree of the graph and then successively improves it by deletions and additions of lines. The method does not necessarily produce optimal solutions, but good results are reported.

For discussion of the problem of Hamilton mentioned in § 10, see

O. Ore, "Theory of Graphs," *Amer. Math. Soc., Colloquium Publ.* **38**, 1962.

W. W. Rouse Ball, *Mathematical Recreations and Essays*, MacMillan, New York, 1947.

J. Lederberg, "Hamilton Circuits of Convex Trivalent Polyhedra," *Amer. Math. Monthly*, **74** (1967) 522–527.

I. Pohl, "A Method for Finding Hamilton Paths and Knight's Tours," *Comm. ACM* **10** (1967) 446–449.

S. M. Roberts and B. Flores, "Systematic Generation of Hamiltonian Circuits," *Comm. ACM* **9** (1966) 690–694.

For a panoramic view of contemporary developments in artificial intelligence, see

N. L. Collins and D. Michie (eds.), *Machine Intelligence 1*, American Elsevier, New York, 1967.

E. Dale and D. Michie (eds.), *Machine Intelligence 2*, American Elsevier, New York, 1968.

D. Michie ed., *Machine Intelligence 3*, American Elsevier, New York, 1968.

G. W. Ernst and A. Newell, "A Case Study in Generality and Problem Solving," *ACM Monographs Series*, 1968.

These contain discussions of the particular puzzles we have considered using various approximate policies. These approximate policies are examples of "heuristic programming" and are particularly useful for high dimensional processes.

Author Index

A

Abramson, P., 96

B

Ball, W.W. Rouse, 138, 170, 211, 231, 237
Bartlett, R.E., 96
Bellman, R., 24, 47, 96, 99, 122, 171, 194, 210, 212, 229, 230, 231, 234, 235, 236
Bellmore, M., 236
Bentley, D.L., 98, 140, 141
Berge, C., 37, 47
Bock, F., 96
Boesch, F.T., 48
Brown, J.R., 234

C

Cameron, S., 96
Carroll, Lewis, 170
Cavalli-Sforza, L.L., 38
Clark, C.E., 48
Collins, N.L., 237
Cooke, K.L., 120, 140, 141, 212, 230
Courant, R., 38
Court, N.A., 173, 195
Coxeter, H.S.M., 38

D

Dale, E., 237
Dantzig, G.B., 120, 121
Davidson, D., 141
Detrick, P., 212
Dijkstra, E.W., 94

Dobbie, J.M., 236
Dreyfus, S., 24, 47, 94, 95, 99, 120, 121, 210
Dudeney, H.E., 139, 172, 232

E

Edwards, A.W.F., 38
Ernst, G.S., 237
Euler, Leonhard, 36, 48

F

Farbey, B.A., 99, 121
Faure, R., 47, 171, 236
Fazar, W., 48
Floyd, R.W., 121
Fraley, R., 212
Fulkerson, D.R., 48, 236

G

Gardner, M., 170, 173
Gaunt, S., 236
Gilbert, E.N., 38
Gluss, B., 235
Golomb, S.W., 233
Graham, L.A., 172, 231
Gruon, R., 47

H

Haggerty, J.B., 211
Hakimi, S.L., 140
Halsey, E., 120
Hamilton, W.R., 220
Hammersley, J.M., 38, 235, 236
Hanan, M., 38

Subject Index

A

Acceleration of convergence, 66, 115
Accessibility lists, 108, 111
  generation of, 160
Adjacency matrix, 83, 154
  powers of, 155
Aitken's delta-squared process, 66
Algorithm, 22, 54–61, *see also* Labelling
  algorithms
  deterministic, 57
  theory of algorithms, 54
  tree-building, 121
Allocation of resources, 114
Arc, 37
Array of times, 16
  enlarged, 20
Artificial intelligence, 228, 237
Assignment problem, 221–226

B

Bandy-ball, game of, 138
Best current estimates, *see* Acceleration of
  convergence
Billiard ball solution, 187
  use in generating Sawyer graph, 188
Branch, 7
Branch and bound method, 225, 236
Branching process, 226

C

Calculus, 22, 52, 63
Cannibals and missionaries puzzle, 196–212
Cascade algorithm, 99

Chess, 232, 233, 234
Chinese fifteen puzzle, 168
Chromatic number, 38
Circuit, 5
Combinatorial problems, 34
Commuting problem, 2, *see also* Shortest
  route problem
  solution by enumeration, 4–6
Complementary matrix, 84
Computer program
  for pouring puzzle, 148, 160, 164
  for routing problem, 82, 84, 109
  for square root, 67
  for travelling salesman problem, 219–221
Computers
  analog, 95, 192
  digital, 192
  heuristic programming, 228, 237
  parallelization, 54, 221
  programming, 57
  use in constructing state graphs, 192
Congruences, 186
Connection matrix, 83, 111, 154
Connectivity matrix, 83
Convergence, 51, 65, 126, 129, 132, *see also*
  Acceleration of convergence
Cost matrix, 152–154, 160, 206
  modified, 165
Critical path, 43–46

D

Defective coin puzzle, 172, 235
Deviation, 95
Difficult crossing puzzles, 196–212
Dijkstra's algorithm, 94, 137
Dimensionality difficulty, 101
Distance between states, 162–166, 204

243

Mathematics in Science and Engineering

A Series of Monographs and Textbooks

Edited by RICHARD BELLMAN, *University of Southern California*

1. T. Y. Thomas. Concepts from Tensor Analysis and Differential Geometry. Second Edition. 1965

2. T. Y. Thomas. Plastic Flow and Fracture in Solids. 1961

3. R. Aris. The Optimal Design of Chemical Reactors: A Study in Dynamic Programming. 1961

4. J. LaSalle and S. Lefschetz. Stability by by Liapunov's Direct Method with Applications. 1961

5. G. Leitmann (ed.). Optimization Techniques: With Applications to Aerospace Systems. 1962

6. R. Bellman and K. L. Cooke. Differential-Difference Equations. 1963

7. F. A. Haight. Mathematical Theories of Traffic Flow. 1963

8. F. V. Atkinson. Discrete and Continuous Boundary Problems. 1964

9. A. Jeffrey and T. Taniuti. Non-Linear Wave Propagation: With Applications to Physics and Magnetohydrodynamics. 1964

10. J. T. Tou. Optimum Design of Digital Control Systems. 1963.

11. H. Flanders. Differential Forms: With Applications to the Physical Sciences. 1963

12. S. M. Roberts. Dynamic Programming in Chemical Engineering and Process Control. 1964

13. S. Lefschetz. Stability of Nonlinear Control Systems. 1965

14. D. N. Chorafas. Systems and Simulation. 1965

15. A. A. Pervozvanskii. Random Processes in Nonlinear Control Systems. 1965

16. M. C. Pease, III. Methods of Matrix Algebra. 1965

17. V. E. Benes. Mathematical Theory of Connecting Networks and Telephone Traffic. 1965

18. W. F. Ames. Nonlinear Partial Differential Equations in Engineering. 1965

19. J. Aczel. Lectures on Functional Equations and Their Applications. 1966

20. R. E. Murphy. Adaptive Processes in Economic Systems. 1965

21. S. E. Dreyfus. Dynamic Programming and the Calculus of Variations. 1965

22. A. A. Fel'dbaum. Optimal Control Systems. 1965

23. A. Halanay. Differential Equations: Stability, Oscillations, Time Lags. 1966

24. M. N. Oguztoreli. Time-Lag Control Systems. 1966

25. D. Sworder. Optimal Adaptive Control Systems. 1966

26. M. Ash. Optimal Shutdown Control of Nuclear Reactors. 1966

27. D. N. Chorafas. Control System Functions and Programming Approaches (In Two Volumes). 1966

28. N. P. Erugin. Linear Systems of Ordinary Differential Equations. 1966

29. S. Marcus. Algebraic Linguistics; Analytical Models. 1967

30. A. M. Liapunov. Stability of Motion. 1966

31. G. Leitmann (ed.). Topics in Optimization. 1967

32. M. Aoki. Optimization of Stochastic Systems. 1967

33. H. J. Kushner. Stochastic Stability and control. 1967

34. M. Urabe. Nonlinear Autonomous Oscillations. 1967

35. F. Calogero. Variable Phase Approach to Potential Scattering. 1967

36. A. Kaufmann. Graphs, Dynamic Programming, and Finite Games. 1967

37. A. Kaufmann and R. Cruon. Dynamic Programming: Sequential Scientific Management. 1967

38. J. H. Ahlberg, E. N. Nilson, and J. L. Walsh. The Theory of Splines and Their Applications. 1967

39. Y. Sawaragi, Y. Sunahara, and T. Nakamizo. Statistical Decision Theory in Adaptive Control Systems. 1967

40. R. Bellman. Introduction to the Mathematical Theory of Control Processes Volume I. 1967 (Volumes II and III in preparation)

41. E. S. Lee. Quasilinearization and Invariant Imbedding. 1968

42. W. Ames. Nonlinear Ordinary Differential Equations in Transport Processes. 1968

43. W. Miller, Jr. Lie Theory and Special Functions. 1968

44. P. B. Bailey, L. F. Shampine, and P. E. Waltman. Nonlinear Two Point Boundary Value Problems. 1968.

45. Iu. P. Petrov. Variational Methods in Optimum Control Theory. 1968

46. O. A. Ladyzhenskaya and N. N. Ural'tseva. Linear and Quasilinear Elliptic Equations. 1968

47. A. Kaufmann and R. Faure. Introduction to Operations Research. 1968

48. C. A. Swanson. Comparison and Oscillation Theory of Linear Differential Equations. 1968

49. R. Hermann. Differential Geometry and the Calculus of Variations. 1968

50. N. K. Jaiswal. Priority Queues. 1968

51. H. Nikaido. Convex Structures and Economic Theory. 1968

52. K. S. Fu. Sequential Methods in Pattern Recognition and Machine Learning. 1968

53. Y. L. Luke. The Special Functions and Their Approximations (In Two Volumes). 1969

54. R. P. Gilbert. Function Theoretic Methods in Partial Differential Equations. 1969

55. V. Lakshmikantham and S. Leela. Differential and Integral Inequalities (In Two Volumes). 1969

56. S. H. Hermes and J. P. LaSalle. Functional Analysis and Time Optimal Control. 1969.

57. M. Iri. Network Flow, Transportation, and Scheduling: Theory and Algorithms. 1969

58. A. Blaquiere, F. Gerard, and G. Leitmann. Quantitative and Qualitative Games. 1969

59. P. L. Falb and J. L. de Jong. Successive Approximation Methods in Control and Oscillation Theory. 1969

60. G. Rosen. Formulations of Classical and Quantum Dynamical Theory. 1969

61. R. Bellman. Methods of Nonlinear Analysis, Volume I. 1970

62. R. Bellman, K. L. Cooke, and J. A. Lockett. Algorithms, Graphs, and Computers. 1970

In preparation

A. H. Jazwinski. Stochastic Processes and Filtering Theory

S. R. McReynolds and P. Dyer. The Computation and Theory of Optimal Control

J. M. Mendel and K. S. Fu. Adaptive, Learning, and Pattern Recognition Systems: Theory and Applications

E. J. Beltrami. Methods of Nonlinear Analysis and Optimization

H. H. Happ. The Theory of Network Diakoptics

M. Mesarovic, D. Macko, and Y. Takahara. Theory of Hierarchical Multilevel Systems